the mystery of matter

the mystery
of matter

exploring the elements of
a metaphysics of process

tony equale

Published 2010 by IED Press
Printed in the United States of America
10 9 8 7 6 5 4 3 2 1

       Library of Congress Cataloging-in-Publication Data

Equale, Tony.
  The mystery of matter : exploring the elements of a metaphysics of
process / Tony Equale.
      p. cm.
  Includes index.
  ISBN 978-1-933567-34-1 (cloth) -- ISBN 978-1-933567-35-8 (pbk.)
  1. Matter. 2. Force and energy. 3. Communitarianism. I. Institute for
Economic Democracy. II. Title.
  BD652.E78 2010
  117--dc22
                              2010048608

Dedicated to the memory of
George Fogarty

... there is no definite concept of body. Rather there is a material world, the properties of which are to be discovered, with no a priori demarcation of what will count as "body." The mind body problem, therefore, cannot even be formulated. The problem cannot be solved because there is no clear way to state it. Unless someone proposes a definite concept of body, we cannot ask if some phenomena exceed its bounds.

Noam Chomsky
Language and Problems of Knowledge, 1988

contents

part 3   **existence**

appendices, *(266-315)*

index

the mystery of matter

preface

The need for this book arose out of certain *lacunae* in an earlier volume, *An Unknown God,* published in 2008. The reflections of that book were directed at specifically religious issues defined and polemicized by the doctrinaire posture of Roman Catholic Christianity. Chapter 4 of that religious *critique* was an early effort to flesh out some of the philosophical assumptions that underlay my perspective. It represented the first attempts to articulate what was for me a new metaphysical view-of-the-world. It was a sketch done in broad strokes which left justifications unestablished and wider significance unexplored.

The point of this book, then, is to turn full attention to those unexplored assumptions, examine them from a strictly philosophical point of view — state them clearly, analyze them, ground them, elucidate the relationships among them, and identify some of the conclusions and corollaries that might flow from them — in search of a coherent synthetic view of the world. I am aware that this represents nothing less than a new discipline — what I call a "cosmo-ontology" — to replace traditional metaphysics. I would have preferred a less ambitious project, but given the unworkable condition of what I inherited from my philosophical tradition, it became unavoidable for me. I may fail at my intended goal, but it won't be for lack of scope. That the results at this point may not be all they someday might be, is due to the vastness of the undertaking ... and the half-vastness of the one attempting it.

I call this book *the Mystery of Matter.* Here, the word "mystery" is meant to translate the Greek word, *mysterion.* In this sense matter is not a puzzle to be solved but a *locus* where the numinous resides and reveals itself. Despite these hieratic overtones, I want to state clearly that religion is a subordinate topic in this study. My primary interest is philosophical. That doesn't mean religion is excluded; religion and philosophy remain eternally entwined in the West. Religion has traditionally related to the same basic questions as philosophy and often even used philosophical descriptors and argumentation in constructing its doctrines.

But religion is not philosophy. Because it is committed to a vision and a way of life, religion naturally resists any questioning of its fundamental expressions, even when they are, at root, philosophical and their current credibility questionable to the point of absurdity. Philosophers, on the other hand, have no such commitments, or shouldn't. Hence philosophy can afford seriously to rethink the traditional understanding of its basic premises without losing self-identity. Philosophy can engage in raw, open-ended enquiry; religion cannot.

What follows is the attempt to initiate just such an enquiry ... one that has been going on in my mind for many years.

a dialog with my mentors

This book ends with some relatively dense appendices that represent my attempt to dialog with people who have wrestled in the past with the questions I take up in this study. It may seem to some readers that there is a pretentiousness in these essays of assessment — that someone like myself should presume to go "toe-to-toe" with these giants when I should rather aspire to sit at their feet. No one is more conscious than I am of my inadequacies in this regard. Hard work and conscientious research cannot compete with the superlative abilities of those historical figures whose ashes I dare to disturb with my probing. If I challenge them, it is only because the accumulated knowledge of the human family has moved beyond what they could have known. But first I want to understand what they thought.

Ludwig Wittgenstein said that it was a matter of complete indifference to him whether his thoughts had been anticipated by someone else. Perhaps it is because I do not fly at his altitudes that I cannot say that. It's a matter of great importance to me that these people with whom I feel privileged to dialog, saw and confronted the same issues that perplex me now. I do not believe these questions are the intellectual games of an academic elite who reside at the pinnacle of a cloud-obscured Olympus, gods to the rest of us. None of us are gods. We are all ordinary people. I would appeal to those who declare that these matters are only pastimes for the professionals, please reconsider. I consider this effort the reappropriation of an ancient enquiry that belongs to the human family, carried forward in every age.

It is our right, if not our obligation as human beings, to try to understand the raucous spellbinding reality into which we were involuntarily plunged at birth. In my case that was 71 years ago; it wrenched me from a silent darkness to which I will shortly return. Forgive me if I am not entirely reconciled to this state of affairs. ... I want to *know what in hell this thing is — **existence** —* that has disturbed my sleep with dreams of endless life*!*

reflection and time

This is a work of philosophy. That means it is focused on thinking things through. It's not that there is no new information offered here, but for the most part you already know the "facts" that will be brought up for examination. What the book does is to invite you to "philosophize" — actively re-perceive, re-analyze and reinterpret our material universe, whose familiarity may have obscured its mystery and blunted its revelatory power. Thinking things through takes time. You *cannot skim* this book.

Take the time. It's worth it.

Willis, VA,
May 2010

acknowledgements

I would like to express my thanks to Zenya Wild and Sheyla Hirshon for their editorial criticisms and suggestions. They have significantly improved the readability of my prose. Frank Lawlor and Terry Sissons helped improve the scientific accuracy and offered important suggestions for clarity. I am very grateful to them. Any factual errors that remain are my responsibility alone. My gratitude also goes to Dick Harding for his commentary on certain sections.

I am grateful to all those who have dialogued with me over the years about issues addressed in this book. In this regard I would like to acknowledge the various degrees of fertile resistance offered by Frank Lawlor, Terry Sissons, Joseph O'Leary, Phil Berryman, Joe Walsh, Joe Holland, Barney Rooney, John Hyland, Bill Duncan, Bob Smith and John Denniston. Thank you, friends. It may not have always been the most congenial and gratifying, but it's all grist for the mill.

introduction

I had a metaphysics professor who used to say, with mock solemnity, "is-ness is our business." It always got a chuckle. It was a humorous attempt on his part to neutralize some of the scorn we students felt about the subject. Whatever his intentions, the remark served to crystall-ize our antipathy for metaphysics in those few words, which became syn-onymous with "waste of time."

I've since changed my mind. The questions addressed by metaphys-ics are important to me. I consider these reflections a metaphysical project, and given my earlier reaction perhaps I should explain why I want to revisit issues that were once the object of such derision. You, I'm sure, have your own take on the matter. What you will be reading, then, are my grapplings with the difficulties the ancient solutions created for me. Per-haps watching me as I rework my view of the world will help you re-exa-mine your own.

The name "metaphysics" is unfortunate. It sounds forbidding — like we should leave it to the professors — like it's none of our business. Most of us do our metaphysics informally and subconsciously; and almost cer-tainly we don't call it "metaphysics." It is basic to everything we do. It is *precisely* the business of "is-ness" — *existence* — that things are present, here with us, and what that means.

Ordinarily, existence is simply taken for granted. Everything we know, do, plan, say or imagine presumes *presence*. We are here. It's because we, other people and the things around us are here all together at the same time and in the same place, that we can work and play, achieve and accomplish, pull up and tear down, build and plant. Existence is the ne-cessary backdrop for everything. It's the air we breathe.

And yet my *being-here* has no guarantee. There was a time when I wasn't here. And there will be a time when I won't be. This *existence* which is necessary for everything I am and do will disappear someday, for me, and for all of us.

"So what?" you might say, we are born and we die. These facts are already part of the background we live with, and seem to have little impact on the flow of daily life. In my case, and perhaps with everyone at some time or another, something happened ... the evanescence of presence emerged from the background and the humdrum of everyday life stopped. The question suddenly became all-consuming: what am "I" if my existence can be taken away? Can I stop this from happening to me, or to those I love? If everything I am or do depends on my being-here, who is this "I," ... and what is this *existence*?

Existence. Why is it here at all? Why "pull up and tear down, build and plant" — why do anything since everything disappears? And this feeling I have, that the world is a wonderful place, ... my home, that I "belong" here ... am I wrong? Every living thing dies. Why am I so shocked to learn that I will die? Where do these feelings come from?

So, from a background taken-for-granted, like the ancient hills on the horizon, always there but never noticed, never questioned, *existence* becomes a question in itself. Hence the business of metaphysics.

metaphysics and religion

Our questions about existence are answered for us, typically, by our religion. Religion's "truths" are meant to explain life and death; they have grounded our sense of the sacred, and justified the feeling of being-at-home. They were part of the horizon of life, public property, shared and common.

But for some, like myself, these common answers are no longer satisfying. They have been discredited by their association with an ancient scientific view-of-the-world that is now obsolete. Modern science describes the universe very differently from the way it was imagined in the ancient Mediterranean world when our religious beliefs were first developed. The ancient world-view is no longer credible — it contradicts reality as we know it — and that fatally affects the religions tied to it, along with our sense of belonging. For those of us accustomed to the traditional world-view, we find ourselves suddenly in a room without air, unable to breathe The whole business of "is-ness" stops being a joke and a waste of time. The "lost horizon" becomes more than a mystery that defies solution; it turns into a wound that will not heal.

Those who manifest symptoms of this existential malaise may have recourse to psycho-therapists who treat our unease as a *disease,* a failure to "adjust to reality." The prescribed adjustment can only mean accepting the "traditional" answers to the big questions, or denying the questions altogether. You have to believe that *you* are the problem, either because you refuse to accept those answers, or because you refuse to call the questions themselves a symptom of your sickness. Those of us who cannot accept those options continue to lead lives of desperation — some not so quiet. The sense of the sacred, a fact of human experience, once had an explanation that sheltered us and gave direction and meaning to life. That interpretation exercises a clamorous nostalgia. Our sense of the sacred does not go away. But now we don't know what it means. We do not cease yearning for home. Aren't our out-of-date religions to blame for this?

No. I submit for your consideration: metaphysics, with its implied "scientific" view of the world, is at the root of the problem. Religion presupposes a metaphysics; it does not create one. All religions are based on some view of the world, however implied and unarticulated. So, as *the primary interpretation of verified facts,* metaphysics is more fundamental than religion. And many religions tacitly contain a metaphysics disguised as divine revelation. In any case, as metaphysics changes, the meaning of the "doctrines" of religion will change.

So I confine myself to metaphysics. The metaphors of religion will conform themselves as they may; I will refer to them and comment, but they are not my primary concern in these reflections.

traditional metaphysics

Those that insist on making metaphysics their business sooner or later pass through the museums and mausoleums of traditional philosophy — in spite of the daunting convolutions of language and ideas enshrined there. Generally, as I've suggested, the enquiry about existence begins with questions about death, but the associated emotions — the fear and trembling — are subdued and sublimated. "To be or not to be," for the traditional philosopher, loses its drama but retains its structure. Some people call the enterprise "ontology," the study (*logos*) of being (*ontos*). Anguish is transposed into rational pursuit and the question becomes a quasi-scientific study of "being" as opposed to "non-being."

Some claim this "ontological approach" is off the mark — exactly the source of the problem. I might not entirely disagree, but I don't believe simply rejecting philosophy is the solution. In the following pages, however, I will follow clues that lead rather to a **cosmo-ontology** — a disciplined methodology that combines metaphysical considerations with physical ones (not unlike the earliest Greek philosophers) about the structure of our universe and the living, sentient, intelligent entities it has spawned. I am searching for a unified answer to the questions about existence and the view of the world that it sustains.

It's hard for me now to believe that at one time I subscribed to the ancient traditional vision despite its being at such variance with the scientific world-view of my times. I was always uncomfortable with that, but I was young and overawed by classical philosophy's perennial prestige. I'm an old man now. Perhaps it's because I am now even older than Plato, Aristotle, and Aquinas when they wrote that I'm no longer overawed by them — Aquinas was only 49 when he died. Like all young people, they shared the certainties of their times. Their certainties are not ours, however. And, whether we like to admit it or not, we have our own. Our view of the world has changed. That change contains an implicit metaphysics that cannot be ignored. This study proposes to make it explicit.

culture and thought

At the base of our personalities lies a vision, *an idea*. Not that we are not our bodies, nor that this idea is separate from us, but that our mind-body unity is so complete that we become "who we think we are."

"Just who do you think you are?" A familiar colloquialism that evokes the commonly recognized social insight that human behavior is determined by personal "definitions," i.e., notions and ideas about who and what we are.

This "thinking" defines us all. It is our culture. The supreme adaptability of the human organism — making it possible for billions of people to speak mutually unfamiliar languages, live in vastly different climatic conditions, feel and obey contrary social obligations, be absolutely convinced about contradictory "truths" — puts the mind-body phenomenon on dramatic display. I once stayed with an indigenous people who ate fried wasps. Those people had bodies that were no different from mine. How could they relish something that was so unthinkable to me? These phenomena reflect the determinative presence of our conscious intelligence pervading every facet of our organism. Our cultures, ultimately, can be translated into *ideas*. And these ideas shape what our bodies feel. Our

bodies are diaphanous; they are thoroughly suffused with cultural imperatives — *ideas*. We are what we think we are.

Now, what we think we are is, in the first instance, an inheritance. It's
what *our society* tells us we are. The social group carries the cultural programming that informs its many individual members what is good and
what is bad, what they should and shouldn't do. And it's clear that by pre-
scribing behavior, our society is also telling us, ultimately, what life is all
about. For, embedded in these preferences, at depths not easily accessible, is a culture's vision, its *idea of the meaning of existence*, a view of the
world — if you will, a metaphysics.

a dialog with culture

My interest in these reflections is precisely to explore the meaning of
existence. This implies a robust dialog with culture. I'm not the first to
realize this. From ancient times questioners have dared to drag the mute
imperatives of culture out of their millennial hiding places in order to articulate them as clear and distinct ideas for the purpose of rational evaluation.
This is critically important. Given the thorough penetration of the individual by the culture, the first step must be the clarification of what lies buried
in symbolic expression and unchallenged assumption. Unfortunately, the
culture's power to introject its vision deep into the psyche of the individual
has not always been fully appreciated. Philosophers have not always
screened their own ideas for the lurking residue of cultural values before
presenting them, with the result that they ended up simply repeating society's preferences in other terms. This is a warning for us. This is not a
facile endeavor.

There's more to it than accurate analysis. The culture's ideas dwell at
depths far below the strata occupied by other ideas. Because they inform
us "who we think we are," cultural values make a lasting, possibly indeli-
ble, imprint ... and our very apparatus of enquiry does not escape that
branding. The culture, in other words, provides the very prism through
which its critics must view its demands and propose alternatives. This
helps explain why the savants may not always be able to objectify these
structures or examine them accurately. The culture is the instrument of its
own criticism.

Cultural preferences offer a fierce resistance to change, and often
even to clear recognition and articulation. They are the horizon into which
all other ideas, if they are to survive, must integrate. These foundation

stones may no longer be identifiable as to source, authorship and original intention, and achieve over time a canonical status impervious to criticism. Re-inventing "who we think we are" is no easy task. It's like trying to dock an ocean liner. The inertia of such a huge body in motion resists alteration.

Nevertheless, most of these cultural notions are resolvable into ideas which were originally, in many cases, the theories of past social analysts and critics — the philosopher-scientists (or culinary geniuses) of an earlier era. This is very significant in our case. Much of what we know as Western culture is attributable to ancestors who long, long ago, challenged an inheritance that they could no longer accept.

the Greeks

The seminal event of this type in the West happened twenty-five centuries ago in the land that we now call Greece and amounted to a cultural revolution. Those people dared to question the accepted wisdom — religious, moral and political — of their own ancient traditions. Historians conjecture about the causes of this unique upheaval. Was it the adolescent antics of the roguish Mediterranean gods that this morally mature people would no longer tolerate? Or, was it the exploitations of the landed aristocracy that drove the populace to question their fatal obeisance before the powerful and to look for political alternatives? The background of this movement may not be clear, but its historical impact on future generations is beyond dispute: the radical thinking that started among the ancient Greeks produced convictions that are with us to this day. They define us. We who have inherited their ideas now speak a multitude of languages and are diffused throughout the lands of the Western world. We no longer remember where those ideas came from, or why people thought they were important. Now they are simply "the way things are." They stand as an absolute horizon against which we view the terrain of reality. Our religions were born in that era. They grew like trees reaching for the light refracted through the prisms of those ideas.

So we are molded by our culture and its vision of the world, much of which was elaborated 2500 years ago. Generally we are unaware of these prejudices and we believe that our thoughts and actions are the result of rational evaluations generated exclusively by current events and evidence. It would never occur to us that what we think and how we decide to act owes its character to an ancient view of the world that is contrary, if not contradictory, to the accepted wisdom of our times.

An essential part of the reflections that follow, therefore, will entail examining our intellectual legacy so we can decide where we stand in its regard. Where did these ideas, and the obligations and expectations associated with them, come from? What kind of world did they imagine and try to explain? What problems did they propose to solve? Perhaps we can discuss them freely, as the ancient Greeks did. Aware of their bedrock status in our lives, however, we are admittedly nervous about sitting in judgment on these ancient notions. We realize that to challenge tradition is to invite reactions of seismic proportions. What if we should find ourselves, no less than the Greeks before us, driven to alter the fundamental operating system that supports the programs by which we live?

ideas and the world of spirit

In our culture, the existence of *spirit* may very well be the *core idea* that explains the entire cultural construct we inherited from our ancient ancestors. That belief created a dualism — that there are *two distinct kinds* of reality in the universe, matter and spirit — that we live with to this day.

The Greeks were awestruck by the power of ideas. They thought ideas transcended time and space, and on that basis they became convinced that *ideas and mind* were examples of a separate genus of being different from matter that they called *pneuma,* "spirit." It made humans different not only in degree *but in kind* from every other life-form on the planet including those like the chimpanzees who were our ancestors. It was a theory that split reality in two. It will be one of the tasks of this book to evaluate that hypothesis in the light of the new knowledge of the universe that our science has given us.

The prime instance of spirit within our traditional world-view is the belief in the separate immortal human "soul." It's a belief that impacts our lives well beyond the obvious function of alleviating anxiety about death. For belief in the "immortal soul" is also the metaphysical foundation of the characteristic *individualism* that sets western culture so dramatically apart from others. It gave rise to the peculiar shape of our social and economic structures, power projections, political charters and religious ideologies. The connection here is as intense and inescapable as it is simple and straightforward. We westerners are who we think we are, with the rights, privileges and obligations we think we have, because since ancient times we have believed that each one of us is a unique spiritual entity, a "person," that has a "soul" independent of the body, that will live forever.

The Greeks who elaborated this *idea,* this theory, long before the advent of Christianity, believed spirit and its emanations were the only "fully real" thing in the material universe. They invested spirit alone, and therefore individual human beings alone, with all respect and value. Everything else — made of "mere matter" — was inferior. They went further, in fact, and said that spirit was *divine*, and considered it *sacred*. In equal measure, matter was demeaned. Matter came to be considered a mere foil for the real reality of spirit and ideas, literally for some, "non-being," and the very essence of the *profane*.

Spirit was the source of being and life; it was sacred. Matter was profane, and responsible for corruption and death. Our material component, the body, was naturally a great liability. Personal development was revealingly called "spirituality." As the term implied, the "flesh" had to be conquered, controlled and eventually shed. The sincere attempt to do this, and in some cases, what appeared to be its actual accomplishment, is critical for understanding our culture.

Without the shackles of the death-bearing material body the human spirit was believed to be as *immortal* as the gods. It gave the individual a transcendent value and eternal destiny. This belief, however, was a two edged sword. On the one hand, it promoted a respect for individuals, their autonomy, freedom and personal achievement that is characteristic of our western culture. But it simultaneously muted the natural instinct for collective action and social cooperation; it provided justification for those who benefited from the suppression of the communitarian dimension. The exploitative stratifications of our societies throughout "christian" history have been a perennial embarrassment — a contradiction to the egalitarianism proclaimed by the message of the man from Nazareth in whose name individual "spirit" was given ultimate value. Quite to the contrary, religion has been used in the West to distract attention from the personal pathologies and social inequities that arose in the wake of dualist individualism ... and thus avoid any redress. "God," it was solemnly proclaimed, supported the program: some should rule and some should obey, depending on the quality of their "spirit" ... known, of course, by their class, wealth, property and power and, as a constant, their gender.

Those ideas — *individual spirit*, life, real being and *the sacred* — mutually reinforced each other to build an integrated cultural complex, with its political and religious derivatives. However inimical to the healthy development of the human individual and society, it was, naturally, like all cultural foundations, resistant to change.

Despite the residence of this *conviction* deep within the western psyche, we are quite capable of examining dualism for what it really is — not a "fact" but an hypothesis, *an idea*. This is important for our work. For once cultural icons like *spirit* are translated back into *ideas*, ideas can meet one another eye to eye. What will emerge from such a confrontation?

mind over matter?

In "modern" times the identification of spirit with "mind," and matter with "body" evolved into a parallel anomaly: a mathematical rationalism, the belief in the almighty power of the *disembodied* human **reason**, with its concepts, its logic and the "principles" that supposedly guarantee it. It was not only used to classify women as an inferior gender, it also justified the subjugation and the exploitation of primitive people, many identified by the color of their skin alone, whose cultures did not invest dry ratiocination with such transcendent value as did ours.

So, I demur. Human consciousness is not simply a calculating machine. The body and its way of apprehending reality is integral to consciousness. It does not calculate, *it interprets and understands*. I embrace the postmodern project that would "deconstruct" the conceits of rationalized knowledge; but I hasten to add: in order to commit myself with even greater passion to the search for *understanding*. *Understanding* is deeper and wider than knowledge and involves the body. This "somatic" dimension is critical to our description of humankind's intimate relationship to the material universe.

evolution

Among the traditional approaches to the existence of "God," arguments from the "order in the universe" were always the most compelling and remained in play through the first two-thirds of the 19th century — well after claims that they were logically convincing were demolished by Hume and Kant. Despite the lack of rigid probity both philosophers had conceded that the complex and intricate beauty of creation evoked the craftsmanship of a Transcendent Intelligence.

That concession ended with Darwin in 1859.

Our scientific understanding of the natural world, developing bit-by-bit since the 14th century, received a defining stamp with Darwin's Theory of Evolution. We now no longer see the structures of nature as "God's Plan." Natural selection has come to be accepted as the real "creator" of the variety of living species. Today we don't believe in the existence of "forms," i.e., blueprints inserted as vital "essences" into an empty dead shell called "matter" by a designing Divinity. With Evolution, a Master-Mind is neither necessary *nor sufficient* to explain cosmogenesis. What there is, is the substrate, *material energy,* evolving life-forms from an inner dynamic *with* which we humans are also assaulted — an irrepressible urge to live, a **conatus sese conservandi** that we experience as our very own selves.[1] At the heart of evolution is a dynamism whose capacity for an apparently infinite creativity, even while our technological culture is determined to control it, we do not understand. What is this dynamism that throbs in all things and we can feel pulsing in our very flesh?

This touches on a central theme of these reflections: The goal is not to know and control ... but *to understand.*

Evolution tells us that process, *becoming,* is what is. Becoming, process, energy, is *existence.* The "Doctrine of God" which has traditionally been a by-product of cosmology and metaphysics — a corollary of our traditional concept of static being, fixed ideas and separate spirit — will be correspondingly affected by this shift. For it will *not* be our traditional concepts of being and spirit, but rather of *becoming and matter's energy* that will guide us to a new understanding of *existence.*

Panta rei. In terms of the most ancient philosophical debates in the Western tradition, this new metaphysics represents a sea-change in fundamental vision — from Parmenides to Heraclitus — in our terms, from static "things" to dynamic process.

the idea of "being" and the idea of "God"

The idea of "God" in our tradition has always been very closely associated with the idea of "being." "Being" is the root idea at the base of the

---

[1] **Conatus sese conservandi** is a latin phrase that means the drive or urge for self-preservation. It was used extensively by the ancient philosophers and closer to our time by Baruch Spinoza who identified it as the nature of being.

western philosophical edifice and the first we will directly examine in this book. And its most important feature for the Greek mind-set was, predictably, that it was an *idea.*

The belief in the transcendent reach of ideas led the Greeks to project onto the things they thought about, characteristics that really weren't there. The way they used the concept of being was a prime example of this fundamental flaw. Transcendental qualities were imputed to "being" that were never directly experienced but rather belonged only to the universal scope of our mental representation of it, *our concept.* We explore this thoroughly and adjust our concepts about being and the information it is claimed to yield accordingly. This will have a profound effect on who we think we are. For the notion of the sacred and how we are related to it has been traditionally controlled by the idea of "being" ... and what we think is sacred is the bridge to our practical choices in morality, politics, environmental responsibility and a host of other areas where values matter.

"being" was "God"

The redefinition of being is not a simple technical exercise, however, for "being" in our tradition is also "God." This ups the ante. I cannot question "being" without realizing that I'm dealing with a religious icon, a personalized hieratic vision, not just another *idea.* "Being" in the West has not only been Aristotle's "unmoved mover," the **centerpiece of the natural order.** "Being" was also traditionally identified as the Jewish-Christian-Islamic "God," an all-powerful anthropomorphized spirit *who resides above and acts* **outside the natural order** to affect the destinies of human beings. This semitic "God" was believed to intervene in human history, sweep aside the powers of this world, perform miracles, reward and punish behavior, enter into "spousal" relationships with individual human beings, and demand nothing less than the obeisance of the whole world. It was belief in that "God" that drove Augustine of Hippo, early in the 5th century of the Common Era to declare the Roman Empire ... and by implication, *all empires* ... "God's" chosen instrument for the salvation of the World.

The traditional Western "God" had a split personality; it was the child of two cultures. According to the Greeks, *ó theos* was **the core of what exists**, the Spiritual Mind whose ideas are the real reality we discern dimly in this world of shadows. According to the Jewish, Christian and Islamic traditions, however, this same "God" was a "person," the only holy thing in a world dominated by false, unholy gods. This "One True God" *stood apart from what exists* and demanded a surrender in obedience to his will alone that sets his followers apart from the rest of what exists — making them sacred amidst the hordes of the profane.

Putting these otherwise contrary traditions together created the peculiar meaning of "sacred" that has characterized the West since the days of Constantine, when Christianity was made the official religion of the Roman Empire. In effect it fused the Semitic duality of sacred and profane with the Greek duality of spirit and matter. "Being" then became hieratically identified with "spirit," and matter as "non-being" was reduced to utter profanity. The religions that were born of that fusion — Diaspora Judaism, Christianity and Islam — all have world-views that to one degree or another, recapitulate those same dualisms.

So clearly, if I challenge the traditional concept of "being," I have to be aware that I am simultaneously challenging the traditional schizoid imagery associated with this "God" — the sacred footer of Western Civilization. It will affect all the prejudices and pre-emptions of our culture. The obligations that our society imposes have been defined by what we believed to be sacred and what we believed to be profane. "Being," — a "Spirit-God," — defined the sacred for us, and our horizon was conformed by it.

Because of the heavy historical overlay that comes with the term, "God," I have avoided it wherever possible. Otherwise, I have put the word in quotation marks in order to emphasize the hypothetical character of its traditional features. We must constantly remind ourselves that "God," no matter how iconic the word has become, is an *idea of ours*.

the sacred

We live in a cultural continuum. That is true both horizontally, along the time-line of history, and vertically, in the interdependence of the elements of culture currently in play. The study undertaken here does not occur in a vacuum, *a la* Descartes, who proposed to "wipe the slate clean" before constructing his geometrical edifice.

The "sacred" is a major feature of our cultural complex. It is not only the ground of the vast and varied institutional phenomenon known as religion stretching back in time for thousands of years, it also pervades all aspects of human activity and endeavor here and now. The "sense of the sacred" is a fact. Our task here is not, in the first instance, to either justify it or deny it, but to understand it and then interpret it in the light of our view of the world.

I have been accused by some of "parachuting in" the "sacred" as if it were something extraneous to human life and its introduction as a topic gratuitous. The assumption behind this attitude seems to be that this study should be starting from ground zero and "build a view of the world" out of some new brick — some new axioms or first principles. If that's what you expect, you will have to do it yourself. The way this study unfolds is different. It starts from where we are; it proposes to understand our legacy on its own terms and then submit it to a *critique* that is directed by our scientifically informed view of the world. There are no institutional loyalties (or hostilities) operating here at the conscious level. We cannot eschew the sacred, both because it is an essential feature of our cultural heritage, and because it is a present phenomenon that impacts the daily choices we make. But we can, and must, analyze it critically and responsibly. This is not a cavalier activity.

As we saw in the case of spirit, many are convinced that our sense of the sacred is fatally wed to the features embedded in the traditional concept of being. I agree and disagree. I agree that the sacred is tied to "being," so if "being" changes, yes, our sense of what is sacred and why it is sacred will also change. By the results of this study, I believe we will never recuperate the "sacred" as we once knew it because "being" as we once conceived it is gone. But in another sense the sacred will not disappear. It cannot, according to the vision promoted here, because **our sense of the sacred is, at root, a resonance of the conatus, our appreciation of our own existence**. This proposition confirms the common opinion of humankind since time immemorial: *it is good to be-here. To be alive or dead is not the same to us.* This is bedrock. Ultimately, it will be our *existence* that will re-interpret the sacred for us. Perhaps it will make us see things differently.

But, of course, as with "God," there is a semantic trap lurking here as well. For the word "sacred" no less than "God" has been given a deep coloration by a long and exclusive association with the "religions of the Book."[2] These religions became the ritual and mythic expression of the

___
[2] Throughout this study I use the term "Book," capitalized, to refer to the Judaeo Christian "Bible." And I do that for two reasons: (1) I want to get away from the words "Bible" or "Scriptures" which

dualisms challenged in this study. They concretized the schizoid spirit-matter and mind-body mis-takes of the ancient world. The attempt to break that link and declare *a new meaning* for the word "sacred," one not determined by a "being" identified with an imaginary "spirit," nor by a "Sheik in the Sky" who "for the sake of his glory" blesses and curses, saves and damns, will almost certainly meet resistance.

I try to deal with the problem of terminology and clarify it as required in each context. But in any case, please be advised: the word "sacred," should it appear, is *not* defined by religion; it is *not* essentialist spiritual "being," and it is *not* opposed to what was traditionally called "profane." It is not identified with the "supernatural" or "salvation" much less by wealth, success or power. *What is sacred is our existence* ... and, naturally, all those things that are its necessary supports.

being is *existence*

Being? Just what is this universal "being" projected by the traditional concept that we criticize so harshly? Isn't it, after all, the same as *existence*, the *presence* we experience and cling to? Not at all. That is the heart of the matter. I am convinced that traditional metaphysics is based on a concept of "being" that does not correspond to reality. It is the result of a dualist imagination, and the metaphysics that is built on it, equally imaginary. "Being" needs to be purified and brought down to earth.

By the old world-view derived from the static traditional "concept of being" we used to think of universal reality as an edifice — an infinitely complicated structure of defined entities whose interrelationships were fixed and stable. Immense as it was, all we had to do was factor it mathematically into its separate terms, from initial necessary infinite cause to ultimate finite contingent effect. In that universe of "things" two plus two equaled four ... always and only four.

But once we redefine being as the dynamic evolving *existence* that we experience in our *presence* here, then the universe no longer appears to be just a chain of "things" linked in horizontal sequence. Rather, like the progressive unfolding of a living organism, it reveals itself to be a vast continuum-in-process, the collective product of a multitude of active partic-

---

have a sacred and unchallengeable connotation in the minds of most westerners; and (2) I want to suggest that a simple "book," which ought to be read as historically conditioned "literature" should not be given a transhistorical authority as the "word of God."

ipants, all receivers and all creators, spinning off spirals of completely new identities on many levels that stack and nest like concentric cylinders. The equations it generates are vertical, diagonal and three-dimensional, with myriads of terms, previously non-existent, emergent, accumulating and integral to the whole.

Two-plus-two, in other words, in the real world, the world of transcendent creative energy we actually live in, yields more than four ... much, much more. This kind of equation needs to be interpreted, not factored, for its calculus is in the service of something else — a huge living surge bearing us forward like a *tsunami*.

existence is sacred

To give away the end of the story: *Existence, material energy* is sacred, the basis of all value. It unfolds into an organic community we call the universe. That community forms the parameters within which our commitments function. *Existence* is the boundary stone and horizon, the ultimate ground and inner compulsion that drives our own determination to survive; from there it injects its communitarian patterns and an intense vitality into our ethics, politics, laws, ecology, aesthetics and mysticism.

Using "moral imperative," as Immanuel Kant did, to ground our sense of care and commitment, reinforced the traditional practice of restricting the sacred, erroneously, to the *human individual*. He could not appreciate the community dimension created by poor, finite, time-bound, clinging *material existence* because his *idea* of "being" with its transcendent, immutable and necessary character, inherited from ancient times, stood in the way. It drove him to look for the sacred in the wrong place — the human individual alone. And so he, like the mainstream of western philosophy, was unable to identify the universal communitarian nature of *existence* and within it, the nature of *the human species*. Humans are the spawn of a single vast community of organic creativity and necessarily recapitulate its characteristics. *Existence is one thing, not two; and that one thing, matter's energy, makes us all one.*

Kant's error was an extension of the foundational errors of the West: the divinization of the *universal idea* and the dualisms of spirit and matter, mind and body, sacred and profane, individual human transcendence and abasement of the material collectivity derived from those dualisms. That he found every rational pathway to "transcendent being" and "immaterial spirit" blocked should have been a clue for him that those ancient cherished notions were simply the products of our imagination.

overview

The search for *understanding*, then, elaborated in the reflections that follow, operates on the assumption that because *matter's energy is existence*, we can understand the living reality in which we are immersed, from which our life was formed, which we are. It seeks to understand *existence*, the source of the sacred — and therefore, ourselves. Toward this end the study covers the following ground:

part 1: being

The first four chapters re-evaluate classical metaphysics, and begin developing an alternative:

— 1: The first chapter contains a critique of the traditional "concept of being," and replaces it with the notion of *existence* (*presence, being-here,*) derived from experience.

— 2: The second challenges the corollaries of "being." (1) It examines the dyad of essence-existence and establishes rather the simple identity of *existence and matter's energy*. It accepts the scientific world-view as a point of departure, and proposes a cosmo-ontology that is consonant with it. (2) It re-evaluates participation-in-being as the metaphysical ground of the "one and the many" as well as the "sacred."

— 3: The third examines the integrative survival strategies of matter's energy — what it does — in a search for what it is. This results in an emphasis on the *integrated function,* what is called *the communitarian strategy,* the basis of all identity, including human "personality." Its verification encourages predictions that higher, more comprehensive *integrated functions* will continue to occur in the future.

— 4: The fourth is a discussion of eschatology.

— Comments and conclusions to part One.

part 2: knowledge and *understanding*

The next six chapters deal with *epistemological* considerations. They elaborate the relationships and distinctions that obtain between "knowledge" and *understanding*:

— 5: The fifth presents a theory of abstraction and human *knowledge* that repudiates the ancient claims of the immaterial transcendence of the rational process and its subject, a separate, exclusively human "spirit." Abstraction is a function of material survival. *Language* is the source of specifically human intelligence.

— 6: The sixth introduces *understanding* as organic (bodily) relatedness. Experience of the "self." The role and evolutionary emergence of abstractive intelligence and its product, *knowledge*.

— 7: Contemplative cognition,

— 8: Interpretation,

— 9: Realization, and

— 10: Recognition (as four conscious operations producing an *understanding* that goes beyond *knowledge*.)

— 11: Metaphoric predication: the language-tool of *understanding*;

— 12: Altruism and society; ethics and politics.

part 3: *existence*

— 13: The final chapter is an interpretation of reality as *existence*, the source of the sense of the sacred. This is the consciously experienced encounter with the unknowable dynamism displayed in the inexplicable perdurance of *existence* in time (survival), the **conatus** (resonance in us) and its derivative, evolutionary development, which proceeds by the communitarian strategy of integrative participation. "Conscious unknowing" is the central paradox for philosophy. It suggests that the very absence of *knowledge* of the process which is *presence*, while it offends rationality, provides *understanding*.

Appendices: There are 14 appendices. These commentaries and illustrations are significantly supportive of the main argument. They also represent important milestones in the development of the *understanding* of *existence* presented in this study. The attempt to integrate them into the body of the text would have obscured the main lines of the presentation with abstruse clutter. They deal with Aristotle, neo-Thomists Lonergan and Rahner, Heidegger, physicists on Universal Oneness, the materiality of "God," Emergence, *Conatus*, Spinoza, a chart of molecular evolution, Lamarck, the social intelligence hypothesis, ethics and politics, Schopenhauer, Bergson, Chomsky.

n our search for ultimate answers about existence, the demand for *knowledge* as traditionally defined will receive little satisfaction. But, once *rational knowing* issuing in fixed conceptualization is demoted from having the right to set the conditions for the only legitimate answers — a different level of conscious embrace is allowed to function. This re-adjustment is central, for I contend it is the towering conceit of a disembodied Cartesian "mind," whose product, the concepts of *"rational knowledge"* — derived from an erroneous belief in the immaterial mental capacities of "spirit"— that has been responsible for preventing *understanding*. It claimed to *know* "being" *rationally and without the body*. And when that knowledge was effectively challenged on its own terms, the enterprise collapsed.

But, *existence* is not our traditional "God-being." The intelligibility of *existence* is not limited to the parameters set by verbalized conceptual knowledge, grounded in the belief of immaterial rational spirit, the idea of "being," the immortal separable soul and the anthropomorphic religious metaphors used to support it. *Existence* is not a "thing," or a manifold entity, much less an idea. Therefore it cannot be *known.* It is a *living energy*, a self-initiated and self-sustained *unfolding material dynamism* that elaborates and sustains us all as material organisms. *Existence* is *understood* through an interpretive realization of the meaning of *the material energy that is **our very selves*** and from there everything material like us that exists. *Existence* is our word for a *becoming* that emerges as us, the enquiring subjects, who are also always the object of the enquiry.

This is the underlying *paradox: to* be both subject and object is *to understand, but it is **not to know** ... and therefore not to control.* To ***not-know***, then, in the precise sense we are proposing in these reflections — not unlike the *un-knowing* offered by the mystics, east and west — is the condition for fully *understanding existence*, the source of our sense of the sacred.

part 1
being

chapter 1
being and *existence*

E xistence is the subject of this entire study. The goal of this first chapter is to define clearly the word *existence.* But in order to do this, we have to separate and distinguish it from another traditional term and concept which is usually taken as a synonym: "being."

1. *existence*

I 'll start by describing what I mean by **existence.**
I want to look at the utterly simple fact that there are things around us. I don't want to decide in advance *what* I think they are, or if or how I think they might be related to or different from one another. I want to focus on the one characteristic they all share, no matter what they might be. And it is the raw fact that *they are all here together.* We are confronted with the simple phenomenon *that* things are *present — they exist.* But what does it mean *to be here*?

Everything that exists *is present,* it's here. And just by *being-here* everything has something in common with everything else: we *are all here* together and *fully present to one another.* That common element I call *existence.* It's because we are here together that we can interact. I can have friends, grow food, build a house, ... because people, soil, plants, boards and nails are all here, fully present to one another and to me at the same time and in the same place. We all *exist*.

Being here is a common trait and a common bond. *Existence* belongs to all of us equally. I experience nothing proportionate or measured about it. Each thing that *is here* is doing the very same as every other thing by being here and has a full measure of *existence*. No one thing is more *present* than any other. By all of us being here together, we realize that there is something very real and intimate to each of us that we all share, and that is *existence*. Whatever *existence* is — and our quest is to understand it — we've all got it, and it's the same for all of us. No one of us *is* more *here* than any other ... no one can be here without it ... we are all fully present.

The next thing we notice is that *existence* is not just some neutral thing we happen to have or not have, like red hair, or a strong back, or even being male or female ... none of which, while we may have our preferences, matter that much. No one *really* cares if they have red hair, but they care very much if they're dead or alive, i.e. here or not. Having *existence* matters. And it matters a lot. I'm very glad that I have *existence* ... but even more than glad, I am deeply invested in *being here* ... when I think about the possibility of *not* being here I become very sad, even morose. The thought of dying holds a certain terror for me. I can't imagine it. I cannot conceive of *not being here*. And even though I know dying is my inevitable destiny, I don't want any part of it. The thing I want is *existence*, and I want it with a passion. I'm aware that I feel driven to do whatever it takes to hold onto it. This urge to self-preservation was called **conatus sese conservandi** by Spinoza, and we will use that term often in these reflections, along with our own phrases, like "the drive to survive," or "the insistence on existence."

Let's go further. From an early age I experienced my *existence* in the circle of my immediate family, who lavished me with a boundless appreciation for the solidity and absolute value of my *being here* with them. Our *existence* as individuals and as a family was attended by an intense and self-sacrificing work ... a sustained maintenance guaranteed by daily hard disciplined labor inside and outside the home. These daily routines were all about *existence, i.e., being-here* and *staying-here. Existence* was taken as an absolute value. It was the reason for everything we did. The supreme desirability of *existence* was not questioned. Moreover, the loss of *existence* was considered the greatest of all tragedies ... of all events the most to be avoided. Sickness, accidents, danger, anything that even vaguely suggested the possibility of the loss of *existence* was immediately prioritized and all available resources were marshaled and brought to bear on it. *Existence was treated as an ultimate and transcendent value.*

Nothing was more important than *existence*. I internalized it as such, and every other living thing I encountered afterward in life confirmed that judgment. Animals and insects cling to *existence* as intensely as I do; plants, more distant but analogous. As I became acquainted with microscopic life forms I saw the same. To continue in *existence* was an innate insuppressible instinct. No one ever questioned, or challenged, that living things will, and **have a right to** do virtually *anything* to survive. And every living thing was the same in that respect. No matter what level of function or complexity they occupied, *they all had a* **conatus sese conservandi.**

2. "being"

*E*xistence is usually understood to be the same thing as **"being."** But it's not. "Being" is the word traditionally used to refer to what exists. It has enjoyed a long history from its beginnings among the ancient Greek philosophers right up to the present day. "Being" in its classic, traditional sense meant everything — all the "things" we see around us, as well as things we may not be able to see, like the human soul, angels and "gods" which were believed to belong to a genus called spirits. It meant all of these things — whatever is.

The Greeks whose worldview we've inherited — mainly Plato and his followers — asked: "what does it mean "**to be**?" Well, clearly, they said, "to be" was *different* for each of the different kinds of existing things out there. A tree existed (and therefore was "present") in a way that was *very different* from the squirrel that lived in it. Since "being" comprehended them both, "being" was considered to be possessed by the tree and the squirrel in a different measure, to a different degree, because the quality of their presence was so clearly different. And the "amount" or "quality" of *"being"* that each type of thing possessed determined *what* that thing was and what it was able to do. Based on this, the Greeks divided "being" into a few broad categories: first, there were non-living things that were totally inert and lifeless, like rocks and water. Then there were plants, which unlike stones and earth were alive but were fixed in place and could not move about, nor see or hear. Next, were the animals who had feelings and mobility but no insight — they could act in marvelous ways that the plants did not but they could not speak or understand. And finally there were we humans who could think and decide, speak and sing, laugh and cry. From the start, being was considered to be possessed *proportionately* by all these different things that were. So from the start, they considered "being" to include what humans knew to be different. *Being was what we **understood** of the things we knew.*

What made things to be *what* they were was not visible on the surface. A donkey had a mouth and two eyes like we did, but it did not understand what it saw and could not speak. So the Greeks said there had to be something invisible, but real and different, inside a human being and a donkey that *explained* why "to be" was different for each of them.

They considered this invisible something to be the most important part of "being." They called it the **"essence"** and realized that by looking very carefully at the thing in front of them, they could list all its characteristics, define it over against other "things," and thus come to *know what* that essence was. So the *idea* of that "thing" they called the *form*, and said it was the same as its **essence**. For a lion, it would be its "lion-ness," all the elements of the definition of a lion, e.g., that it was a mammalian vertebrate, of the feline class of a certain large size that preyed on large animals, traveled in prides of many individuals, etc. It was what gave the lion its form and function, its "reality," its "being." Whether or not this essence actually *existed* as a concrete individual was only a secondary and less significant part of its reality, its "being." What it *was,* was *to be* a lion.

So, when traditional philosophers like Plato used the word "being," this is what they meant. "Being" was a *composite* primarily constructed of that *invisible factor, the essence,* that made a particular thing to be *what* it was and different from everything else. That *essence,* said Plato, is accessible to our probing intellect. If we look carefully at the thing, we can form an *idea* which is the mirror image of its *essence.* So, whether that "thing" was actually there in front of us or not, we could hold the *idea* of it in our minds. The idea, therefore, had a certain reality of its own, and curiously, a most important reality because it made things *to be* what they were and not something else. The fact that the entity in question *actually exists* or not, is the less important component, and almost something accidental to the *invisible idea.* The *essence,* the idea, was what told us *what* the thing *was.* It was the most important part of its "being."

This highlights the abstract nature of the traditional term and notion "being." For traditional metaphysics, the concept "being" *was not empirically derived.* It was the result of a prior analysis that decided "being" had to be *different for different things* and therefore structured of *essence* and *existence.* "Being," they said, had to be possessed *proportionately* by different kinds of "being."

But sometimes the Greeks used the word "to be" to refer to *existence* alone. When they did that, they understood it as if it were, in scholastic terms, *an accident.*[3] In this usage, "being" refers to *existence* as if it were only a potential characteristic,[4] a separate quality which may (or may not) be predicated of a substance.[5]

*[**Author's note**: there is a possible confusion between this less common use of the word ("existence" as the companion of "essence") and "existence" as I intend it, and I want to clarify: for the purposes of this study I will generally use "being" to refer to the composite Greek-scholastic term made up of essence-and-existence, as described above, and "existence" alone to refer to the actual presence of things, without coupling it with essence except when I'm dealing with that traditional dyad directly. It should be clear in context.]*

"being" is different from *existence*

We need to distinguish traditional "being" as just described from what we called *existence, being-here-now,* as we saw in the prior section, by which we mean *existing in the present moment. Existence-as-presence* is a raw datum; "being" is not. *Existence* evokes the experience of simple, uncomplicated here-and-now *presence* that results in a judgment, "this lion *is-here,* now" that precedes any and all "looking," abstractions, analyses, reasonings or insights into the differences between things. It refers to an experience, not, as "being" does, to a concept, or an attribute of an

---

[3] Thomas Aquinas, *Quaestiones Quodlibetales 2*, qu.2, art. 3: "... therefore the actuality, or *existence*, that relates to the question *whether* something is, is an *accidens*." Quoted by Heidegger in *The Basic Problems of Phenomenology*, 1927, 1975, tr Hofstadter 1982, Indiana U.Press, *y* p.91

[4] Kant understood this "existential" usage of the concept of being. He said "being" was not a "predicate" that added anything to a "reality" (by which he meant "essence" or logical "substance"); *actual existence*, for Kant, was a "relation" to the knowing subject. Cf. his Chapter on the Transcendental Ideal in the *Dialectic of Pure Reason*, (A.599; B.627), Anchor 1966, tr Muller, p.401. In this regard see also Heidegger's extensive analysis of Kant's "concept of being" in *The Basic Problems of Phenomenology*, op.cit., p.27ff.

[5] following the traditional nomenclature, a "substance" is composed of *matter and form.* It means a "thing," an individual example of an "essence," and sometimes by abstract extension, the essence itself. "Being," can be an abstract universal and as such is composed of *essence and existence.* But "a being" or "beings" refer to concrete individuals, and mean "substance(s)."

essence. And while all experiences lend themselves to conceptualization, the concepts (and the implications of the concepts) derived from the perception of *presence, being-here-now, existence,* are different and yield different results — they imply a different philosophical understanding of the world — from the traditional, classic concept of "being."

The experience of *presence-in-time* which we are calling *existence,* is initially a datum, a perception, an experience. It is not derived from predication and so it carries with it none of the implications of its use as a secondary composite of "being" dominated by the predicate — the *idea,* the essence. The perception of *existence* is homogeneous in every instance of its occurrence, as we saw. There is *no* experience in the present moment of whatever kind that is *not* like it. In the world of experience, which is the real world, the world we live in, *presence, existence,* accompanies every single event and entity without exception. Nothing is encountered or experienced that does not *exist now,* i.e. that *is not here.* There is no abstract "lion" which may or may not exist; in other words, there is no "potential being." In the real world each and every lion *actually exists. Being-here* in this world of experience, the real world, is not *an accident,* or a mere possibility. *Being-here* is an absolute necessity, in the sense that there is nothing in the real world that *is-here now* without it. *Existence* is all there is. All experience is only and always the experience of *present existence*; it is the experience of what is necessarily *present* because it is being experienced. There is no experience, ever, of anything that *is not here now.* There is no abstract or potential "being" anywhere. Abstract "being," and, as we shall see, its counterpart "non-being," does not exist, except in our minds.

> [**Author's note:** *Because of the universality of the experience of* **existence** *(that everything is as present to us as every other thing), I am inclined to equate existence with the* **conatus***, which is, in similar fashion, found in* **all living things***, across all the phyla and levels of function. This suggests that the phenomenon we have labeled* **conatus** *may be equally functional wherever* **existence** *is found, even among non-living things. This remains to be seen.*

In the real world, then, *existence* is not a potential attribute predicated of a *substance.* It is active, concrete *presence,* and for all we know, it's all there ever was. "*Existence*," in the real world, is not a quality separate from the things that are. It is only encountered as the actual concrete *being-here* of something — *it is the actual presence of what's actually present here and now.* Any concept derived from the experience of *being-*

here, used for purposes of philosophical analysis is universalized from this absolutely unique perception and judgment of experience that applies necessarily to the whole universe of things currently existing and does not share its predicative significance with anything else, and certainly not with "non-being." *Existence* is not a "quality" of something, as the traditional concept of "being" would have it. It is the active actuality of the actually existing something. Without it there is no "something." Apart from *existence* there is nothing.

This starting point is not intended to deny the validity of concepts. There is nothing to prevent us from conceptualizing this particular awareness of *presence*. The *concepts* then, derived from *existence* will speak of *recognizing the similarity of experiences*, not generating an essence or a defining difference ... not attributing an abstract potential, a logical construct, or a quality to another mental construct.

This difference will have its first and possibly most decisive impact on the traditionally accepted **concept of "non-being."** For, in the real world, there is no concept of "non-being" that stands on a par with "being" because "non-being" is never experienced. And it cannot be experienced because there is no "non-being" in the real world. "Non-being" simply does *not exist.* And that means, equally, there is no universal concept of "being" that may be considered parallel to "non-being" or, as we shall see, that derives its "significance" from it.

3. *"being"* and nothingness: the metaphysical question [6]

The traditional focus on the "transcendent" qualities of "being" — that "being" goes beyond the finite phenomena in which it is experienced, and is in fact *infinite* — have been characteristic of western philosophy most particularly since the middle ages, when this feature became the basis of a deeper exploration of the Creator-creature relationship inspired by the Judaeo-Christian concept of "God." [7]   Besides, "being" as understood by the scholastics, was *intrinsically "intelligible,"* as we've seen, because it was primarily characterized by *essence,* the *idea*

---

[6] The argumentation and even some of the text of this section and the next two, was anticipated in chapter 4 of *An Unknown God.* I repeat it to maintain the integrity of the overall presentation.

[7] "God's" Hebrew name, *Yahweh,* was creatively *mis*-taken to mean "I am, Who am" — in a literal, philosophical sense, "Being." It is now commonly acknowledged, however, that *Yahweh* means "*I am who I am*" in effect a refusal to give a "name," which in the cultural context would have meant giving magical power over the god.

that put *purpose* and therefore definability into the things that exist. *Purpose* implies design and a designer. Hence traditional "intelligible being" implied a rational (*idea*-making, purposeful) creative "God" as well as an *immaterial essence* conceivable by us humans *as an idea-with-a-purpose.* The foundational dualism of the Platonic vision is necessarily embedded and implied in all these interdependent notions. Immaterial *idea* was at the core of Greek reality.

But the scholastics took the Greek world-view one step further. Neo-Thomists like Rahner and Lonergan, claiming Thomas as their source, have used the concept of being's *opening to transcendence* to ground their philosophical theology.[8] It's certainly true that Thomas himself focused on the implications of the **concept of "being,"** rather than, as the ancients did, on a *Nous,* a Creative Thinking Mind, to explain the existence of intelligible "things." The concept of "being" was necessarily infinite. "Infinite being," if taken in a *real*, not just logical sense, implied not only the existence of all things but also the existence and character of "God,"[9]

finite *existence*

The finite version of "being" — what we have been calling the *existence* of things — is, however, the only way we actually encounter "being" in the real world. *Being-here-now* is what real things actually *do,* it is quite finite, and *that* is the only "being" we experience. Even though we can *imagine it,* we do not *experience* infinite being

---

[8] See the appendix for a lengthy discussion of this and references.

[9] Since an attribute of "being" was that it had to be "One," it posed a problem. It meant finite being had to somehow "share" God's being. Traditional philosophy called that "participation-in-being." This established the character of what today is called "pan-entheism." We should remember that, hovering in the background of this mediaeval conception, lay the millennial confusion between concepts and reality (the source of Plato's theory of Forms), what the schoolmen debated for two hundred years — the "problem of universals." Until the definitive resolution of that problem in the 14th century, the philosophers *tended to impute extra-mental reality to their concepts* — and the concept of "being" implied "God."

The mystical implications of this line of thought early in the 14th century were met with resistance from the inquisition. Spiritual leaders like Marguerite Porte and Johannes Eckhart were condemned as **pantheists**. The monumental achievements of scholastic thought, culminating a thousand years of development based on the Platonic vision, derived from the implications of a metaphysical participation, a "sharing" of being, became an increasingly marginated esoteric doctrine confined to the monasteries.

Saying that "being" is infinite has been traditional in the West. Of course, if we are referring to the *logical properties* of the "concept," who can argue. But, I disagree that "being" is "transcendent" or infinite beyond the logical characteristics of the concept. I claim and plan to show that there is a *pre-conceptual imagery* that served as source for the concept. It is actually the more basic feature of the traditional western projections about "being" and they have shaped its character definitively. We have already seen some of it in the traditional practice of interpreting the significance of "being" from *what* things are, their *essences*. But there's more: ***"being" was defined by "non-being."*** The analysis we employ here in correcting this misperception reveals how it has prevented us from seeing the reality — *existence* — which actually stands before us.

Let's look at this more closely.

death and *existence*

The motivation behind the metaphysical enterprise has been identified with the human inclination to question existence itself: ***Why is there something rather than nothing?***[10] This is the classic form of the *metaphysical question*. Reasonable as it sounds, it is actually quite extraordinary. Usually no one ever thinks about *existence*. The psychological context that gives rise to this particular question is often associated with some event that "awakens" the questioner out of the "forgetfulness of being,"[11] an alleged metaphysical torpor that is said to accompany daily life. This "wake-up call" simultaneously extracts *existence* from being a background condition, an invisible horizon, and sets it front and center before human consciousness as a direct object of enquiry. Prior to this intentional formulation, *existence* as such is not questioned. It is *not* a normal object of knowledge.

There are many experiences that can provide the occasion for this change of gears, this awakening. The most commonly cited is an encounter with *death* or its equivalent. Such experiences reveal the question in the first instance to be about human existence as we know it. That it is a "rude awakening," a "shock," is explained by its roots in the ***conatus.***

While the metaphysical question, *Why is there something rather than nothing?* is claimed to stand on its own independent of its link to these

---

[10] See Appendix 2: "Heidegger."
[11] This is Heidegger's expression.

psychological origins, I contend, to the contrary, that *the experience of death or its equivalent*, an experience that makes reality seem ephemeral or evanescent, has in fact predominated in the West and given the metaphysical question its traditional character.

Modern existentialism's almost exclusive focus on death as the starting point of authentic philosophy was an outgrowth of that predominance. I believe that the very form of the question, *why is there something rather than nothing,* is considered "natural" only because it is wed to the psychological phenomena that have occasioned it.

I have two preliminary observations. First, we should notice the fact that the question, *Why is there something rather than nothing,* is not ordinary, as we mentioned. It does not arise in the process of daily living. It specifically requires that a previously *unnoticed feature of the cognitive horizon*, "*existence*," change its status from background to foreground and become the focus of investigation and concern. We will come back to this later.

Secondly, we should look at the structure of the question itself. The question doesn't ask, *what is existence? or what does existence do?, or what is it for?*, it asks specifically, Why is there something *rather than nothing?* The ruling notion in the question is "***nothing****.*" By saying, "rather than nothing," the question assumes we know what "nothing" is and that there *ought to be "nothing,"* and registers a certain surprise to find that *there is something.* What can that mean, *there ought to be nothing?* It seems to give a place of primacy to *nothing.* And why does it seem natural to find "being" a revelation and an anomaly? The strange statement, that *nothing ought to be,* should not be glossed over as a mere incongruence. It is, in fact, an utter contradiction. And, upon reflection, we realize it's not a *logical* contradiction (because grammatically the concept "nothing" is no different from any other concept, even the concept of "being"), but it is a *metaphysical* contradiction. For it is obvious that "nothing" cannot *be* in any form or fashion. *There is no such thing as nothing!* ... so you cannot say *it ought to be.*

What is "*nothing?*" I must know what *nothing* is, ... how else could I use it to formulate the most ultimate of questions? But it turns out that *I do not know "nothing."* I have never had any experience of "nothing" from which I could have formed a concept. *My only experience has exclusively been of what exists now.* Furthermore, I cannot claim to derive my concept of "non-being" from "being" (for example, as a "negation" of being) because I am using "non-being" to understand "being" itself. To use each concept to explain the other is a gross circularity. So I have no legitimate source of my concept.

What could be the source of our *idea* of "nothing"? Various pathways have been suggested.[12] Working from the primacy of experience in providing us with ideas, what makes most sense to me is that the idea of "non-being" is imaginatively extrapolated from the *experience of death and birth,* the generation and dissolution characteristic of all living things that we know, and that includes ourselves. We cannot imagine after-death or before-birth, because we cannot imagine *non-experience.* So we call it "non-being." But non-experience is not necessarily non-being.

It's here that I begin to understand how my awareness of the vulnerability of *my life* as an experiencing person has functioned to formulate the terms, notions and structure of the metaphysical question, "*Why is there something rather than nothing?*" Someday sooner or later, like every other living thing, "I" will die, "I" will not *be-here.* But the only thing I really know is that "I" will be neither subject nor object of *my experience.* The important fact (to me) of my *death* (my ceasing to experience), has apparently generated my assumption of the disappearance of "my existence." I have identified *my life,* by which I can only mean *my experiencing self,* with an instance of *existence.* The **conatus** is responsible for this. When *I die,* I am assuming that *"I" will cease to exist. Existence,* therefore, by this definition, can cease. This "ceasing to exist," I submit, is the origin of my concept of "nothingness."

But what if I am mistaken about this "metaphysical" assessment of death, which amounts to a projection, an interpretation, a metaphysical evaluation of my experiencing self? ... and perhaps an exaggeration created by the intense reaction of the **conatus**? What if death does not mean that I *cease to exist.* Let's look at the two possibilities about the continuing *existence* of the entity I call "I."

---

[12] Bergson has an extensive treatment on the possible origins of the concept of "the naught" in his 1904 book, *Creative Evolution,* (tr. Mitchell, 1911) Dover reprint 1998, pp.272-298. He believes it is a false notion, a reification improperly extrapolated from logical negation. Negation is also Hegel's source for the concept; the same is true for Sartre. Most existentialists, however, following Kierkegaard, derive it from the experience of death and claim it is central to the authenticity of the human response to *existence*: the source of ethics and politics.

My *experiencing self* is either something independent of my mortal body or it's not. What if, first, as western culture has traditionally asserted, my *self* is or has a separate *immaterial* "spirit," an immortal soul, which is capable of living *without my body?* Then, when I die, my "spirit" lives on. In this case, death only affects my body, it does not affect my *self* and "I" do not *cease to exist.*

Or, the other alternative, what if my *self* is not independent of my body but is simply the *perception of the individuality of the organism* which I am, the efflorescence or reverberation of the particular combination of sub-atomic particles that make up the elements that comprise my human organism. My *experiencing self*, in this case, is *not* an entity apart from the "parts" that make up my body. When my body "dies" it simply means its particular combination of elements no longer coheres, the inter-relation-ship among my components changes, my *experiencing* vanishes with the loss of organic integrity, but the bits and pieces of which "I" was made go on to become available for use by other things. There is no "ceasing to exist" in this case either, because my *experiencing self was never an independently existing entity to begin with.* Everything "I" was made of continues on in other things just as it had all been part of other things prior to being incorporated into my body. *Nothing has disappeared, nothing has ceased to exist.*

So I realize that in either case, to equate death with *non-existence* or "nothing" was improper, a projection based on an erroneous assessment of the metaphysical significance of my *experiencing self* driven by a repulsion that comes from a **conatus sese conservandi.** The privileged position that I gave to "nothing" in the metaphysical question had no basis in fact. *Therefore I cannot ask it.* I have no reason for saying that "being" ever *ceases to exist* and therefore I cannot say "nothing" *exists* either. Nothing *never existed,* and it most certainly cannot be said that it *ought to exist.* There is no such thing as "nothing." I may still ask metaphysical questions but they cannot be asked in the form we have been examining, *"Why is there something rather than nothing?"*

We may ask, for example, *what does being do?* Or *what is its purpose, or meaning?* Or w*hat is this "I" that means so much to me? What is the **conatus**?* Is there a "God" responsible for existence or is existence an endless presence that contains within itself the source of its dynamic manifestations? I am not restricted from asking any number of metaphysical questions, but to question "being" uniquely from the point of view of an imaginary *nothingness* can only yield imaginary results. It will be a logical tautology, a begging of the question. For the very form of the question itself guarantees the result.

This has been our traditional procedure in the West and it has created "metaphysics" as we have known it. It led to the conception of "being" as *necessary, absolute, unique, infinite, static, fixed and eternal.*

4. the significance of "nothingness"

It is the assumed primordiality of "nothing" as the spoken or unspoken horizon against which the question of *existence* is posed that has definitively determined the salient characteristics of "being" in our western cultural mindset. For in the context of a "non-being" that *ought to exist,* the presence of "being" can only represent an unexpected triumph, a conquest, a choice, a decision, the *active suppression* of what is believed really belongs there ("non-being"), in order to *forcibly introduce* what doesn't belong there ("being"). Two corollary notions immediately insinuate themselves: *intention* and *power.* "Being," in a world where nothingness reigns supreme, has to be *purposely inserted and willfully sustained* with an ultimate metaphysical *power* capable not only of overcoming nothingness itself, but of preserving "being" intact under conditions that could be called "foreign occupation."

The primacy of nothingness is traditional in Western Philosophy. Sartre imagined we live on an isolated strand of "being" surrounded by an "ocean of nothingness." Heidegger claimed *das Nichts* (nothingness) constitutes the ultimate and defining horizon of human life and he speaks of being in terms of a "decision against" and an "overcoming of" nothingness.[13] "Being" that doesn't belong can be expected to be sucked back

---

[13] See below, Appendix "Heidegger — Nothingness and being." "Nothingness" (*das Nichts*) is the theme of a talk Heidegger gave in 1927 called *What is Metaphysics?* The address ended with the formulation of the "metaphysical question" exactly as phrased here, **"why is there something rather than nothing."** Later, in *An Introduction to Metaphysics* (lectures of 1935, published in 1953, Eng tr 1959, Anchor, 1962), H. opens with that same formulation of the question and spends the entire first chapter (42 pages) exploring its implications. He says that the question means *"a ground is sought which will explain the emergence of the essent as an **overcoming** of nothingness. The ground that is now asked for is the ground of the **decision** for the present essent over against nothingness,..."* (p.23, emphases mine). This "overcoming" and "decision" is the basis for his ultimate definition of being as *phüsis,* power, and is the inevitable result of asking the question in that way, as we have observed. And even though H. recognizes the contradictory nature of the very concept of "nothing," he insists on this formulation of the question on the self-justified grounds

into the nothingness that presumably does, unless it is *actively prevented* from doing so by a power that possesses "being" so incontestably that it can overcome and resist the inertial primordiality of nothingness.[14] Thus the metaphysical question, *why is there something rather than nothing,* if allowed to be asked, yields a preliminary answer: *if something is,* **something must be.**[15] Thus is *necessity* introduced as the first feature of "being."

Before commenting on the implications of this preliminary analysis, certain inconsistencies of this line of thought should be noted. The word "nothingness," in the form of a substantive by the use of the suffix "-ness," evokes the sense of a *state* which we recognize is, again, a metaphysical contradiction. In this usage, "nothing" tends to be treated *as if it were an existent,* which is impossible. Observe also that every reference to nothing in the above paragraph — "primordiality," "ought to exist," "belongs there," — implies the same: the words, inappropriately, all evoke the *existence* of nothing. But nothing does not exist. There is no such thing as nothing. We cannot even *think it.* When we try, we immediately couch it *in terms of existence.* Our minds are organic apparatuses evolved by and for *existence* and they are *organically incapable of conceiving non-existence.* And that serves as a corroboration of the contention made here: that the notion of nothing or non-being is a sheer product of our imagination, a fantasy derived from our projections. If it is used as a functional element of analysis it will invalidate any conclusions drawn from it

Once the question, *"Why is there something rather than nothing?"* is permitted to be asked, the preliminary answer, *if something exists, something* **must** *exist,* establishes the defining step — *necessity* — towards the classic characterization of "being" and the creation of traditional west-

---

that, like poetry, truly great philosophy goes beyond the bounds of ordinary "logical" thinking. He offers no other explanation.

[14] Heidegger proceeds from his premises to define being, logically, as "power," in Greek, *phüsis.* While traditional Christian metaphysics used the question to project a Creator responsible for this nothing-conquering power, Heidegger uses the question *existentially* to project a creative *Da-sein* (human being) who necessarily recapitulates being as a creative, productive force. Cf *Phenomenology,* p.101.

[15] Kant's famous dilemma about the Transcendental Ideal (God) — i.e., that necessary being must be attributed to the totality of being but cannot be affirmed of *a* "Necessary being" — is based on exactly this assumption. Op.cit, Chapter V.

ern metaphysics.[16] "Being" that *must exist* is **necessary**. Once this conclusion is accepted, the rest follows with logical inexorability:

Being that is necessary, *has to be*, and therefore it must be **absolute,**
that is, it must be totally independent, relying on nothing but itself for its
existence or powers. If it were not absolute it would have to be dependent
on another, and then it would not be necessary.

It follows, then, that there can only be **one** "being" *qua* being, for two or
more absolutes are a contradiction — they would necessarily condition
each other and effectively cancel out each others' absoluteness.

Such being is also **infinite** in time, space and power, it can have had
no beginning and no end and no limit to its powers over nothingness and
within being or else it would no longer be absolute.

Other qualities also follow. "Being" is absolutely **immutable**, because
movement or change of any kind would imply either an increase or dimi-
nishment of something, which would contradict "being's" absoluteness.
"Being" that is absolutely immutable, of course, because it needs nothing,
already possesses everything it could ever need, is **perfect** (in Aristotle's
terms, has realized all its potential and is **Pure Act**). Such "being" is also
completely **simple**, without parts, divisions or sequences of any kind internal or external, in time or space (implying movement, change, or component elements which would *not* be absolute and necessary), and **immaterial**, because "matter" is of its essence extended, composed and sequential in time ... it is intrinsically **finite** (and as a corollary, also suggests
an absence of being).

Now, the argument continues, if there happen to exist some things that
do *not* fulfill these necessary qualities of "the *idea of* being" — that is,
things that possess "being" but are in fact conditioned, dependent, material, sequential and finite (like everything in our entire universe) — they
must exist *dependant upon* the absolute, and their *existence* must not only
find its ground in the absolute, they must somehow *participate* in the very
existence of the absolute, since, we have seen, all being is **one**. This also
implies that Infinite Being must have a **creating and sustaining power**
with respect to finite being.

---

[16] We must recognize that this was true even for Kant, who, while he denied the logical validity of
deriving "God," The Necessary Entity, from necessary being, *did not ever deny the latter*. That is
because his philosophy was in the mainstream of classic western idealism: he also exclusively
analyzed ideas, and for him, "being" as an idea, was necessary.

They must also be **distinct**. Finite and infinite being are *logical contraries*; they both cannot "be" in exactly the same way. If, furthermore, there exist these different modes of "being" — one infinite and absolute, and the others finite and contingent — then "being" is an **analogous** quality; it resides in different kinds of being in proportionately different ways. There must be different "essences" to account for these differences.

Such are the classic derivatives of "being." They are logically deduced characteristics demanded by the nature of the concept generated against the horizon of nothingness. Since "God" was believed to be "being" itself, these characteristics of the concept were considered "attributes" of "God."

It made us think we knew "God."

5. the *insignificance* of "being"

All this is implied in the apparently simple answer, *If something is, something **must** be,* to the so-called "natural" question, "*Why is there something rather than nothing?*" As we realize, the question itself, directed by a non-existent nothingness, far from natural, was improper. The implications of this impropriety are huge:

Concepts function by separating out and focusing on the salient characteristics of what we know. They identify their objects by specifying *differences* within classifications between one object and another. To know "being," however, is to focus on an object that is *not different from anything*. Everything exists. The only "thing" that "being" can be different from is non-being. But non-being does not exist. So, without contrast to non-being, *being has no specific significance* — it has *no meaning* because it's not different from anything.

This agrees with our short analysis. *Existence,* as an *idea,* tells us *nothing. Existence,* unlike "being," is *not intelligible*; it is simply *a fact.* This is important and its significance will become apparent as our analysis develops. *Existence* tells us nothing because it is not, as "being," subordinated to "essence" the *idea-source* of intelligibility. *Presence* is the same no matter what class or species of "substance" is experienced as *being present,* from stars to starfish, asters to asteroids. The perception of *presence* is absolutely homogeneous and so the concept, far from being analogous, as the Thomists claim for "being," is absolutely **univocal**. *Existence-in-time* does not inform us that in fact things are *not* the same at all, for insofar as they *are-here-now,* they are all doing exactly the same thing. Even the fact of their *congenital impermanence,* that they become something else in time, leaves them existentially unmodified, for whatever they *become* is also only and always *here-now*. All *presence* tells us is that we are immersed in a world of *uniformly present existence*, and "being" is the word we have traditionally used to identify it. But we are seeing that the word "being" has been distorted by its conceptual character and does not properly represent reality. It is, in fact, a *projection.*

The assumption of the "primordiality of non-being" was integral to the judgment that "being" is a *positive quality*, transcendently significant, rather than the *ordinary and insignificant universal condition,* a matter of raw *fact,* the foundation of everything, the only thing there is. It is only the comparison with an imagined non-being that gives "being" its transcendent metaphysical significance in our tradition. Without the backdrop of non-being, an enquiry into "being" is meaningless, for "being" is all there is. So the question, "why is there something rather than nothing," is absurd. Henri Bergson puts it this way:

> "... the question, 'Why does something exist?' is consequently without meaning, a pseudo-problem raised about a pseudo-idea." [17]

Eliminating this absurdity would displace the ontological axiom, "if being is, being **must** be." "Being" would lose its **necessary** status. For, an existence that does not presuppose a conquest over "non-being" is not necessary, *it just is* — and may always have been as some of the early Greeks believed. If we were to erase the *necessary* part of it (i.e., drop the claim that we *know* "being" is necessary), the statement refers only to insignificant *presence* and the axiom would be reduced to an empty redundancy: "if being is, being is." In Lonergan's words, a mere "matter of fact." [18] Unlike "being" which claims to discover **necessity**, *existence* in and of itself gives us no new information at all.

But that's entirely appropriate, for in fact it's *all we know*. The note of **necessity** does not come from our observations of *existence* which is universally contingent, finite, temporary and changing; it comes from the human conceptualizing process. Karl Rahner admits that the **absolute infinity** of being does not derive from our experience of finite being, it is rather, he believes, a transcendent *a priori* that is *projected* by human knowing, the centerpiece of his theory of abstraction. He claims that the

[17] Bergson, *Creative Evolution,* p 296.
[18] See Lonergan, *Insight,* ch.19. He claims mere "matters of fact" do not exist. For him all being is intelligible. See the appendix on Lonergan.

ability to form concepts, i.e., *to universalize*, presupposes a comparison with a universal horizon, which he identifies with the *a priori* "pre-apprehension" of "infinity." Conceptualization itself, in Rahner's view, not only validly reveals "being" to be absolute, it simultaneously reveals "human nature" to be transcendent, that is, immaterial, spirit.[19] I consider all this invalid. To imagine something that I do not experience, and then call *that imaginary projection* a proof of its existence is a vicious circle.

If *existence* is simply *here-now*, an uncontested *presence*, as we are claiming, it leaves open the possibility of the evolutionary elaboration of "finite entities" without the dilemma created by having to explain their provenance from an **infinite**, **absolute**, fully complete "**One**" in "Pure Act." We can say that "things" *evolve* from a *potentiality* within being (as science in fact observes) that does not contradict being's character because we have not pre-defined "being" as "Pure Act," a fixed and completed feature which was only justified by the character of universals or the mental gymnastics necessary to overcome the inertial primordiality of an imagined "non-being." Without "non-being," "being" can be *finite*, which is, again, exactly the way we find it. If being only "is" rather than "must be" then we can say "being" without saying "Pure Act." "Being" in this case, however immensely extended in both time and space, *would **not** be infinite*. That would mean the source and ground of *existence* (what we traditionally called "God") might not be as traditionally defined — static and eternal, finished and complete. If "being" is *not* perfect, it might be *open to development*, and in fact may be actively developing, *in process,* even as we speak, as indeed it is. We have traditionally said that such a possibility would make "being" *finite*, and rejected it for that reason. "Being" taken as *presence, existence,* however, as I am suggesting, *may very well be finite with an infinite potential* for what currently does not exist. This would correlate perfectly with the actual universe-in-process as we find it.[20] It may imply a "source of *existence*" that is not distinct from the finite reality we know — and which includes us.

---

[19] Rahner, *Spirit in the World*, op cit, *this is the main thrust of the book.* Cf. also pp.162 and 175: his few statements about the vacuity of "presence" as opposed to the traditional *universal* concept of "being." See also my appendix here on Rahner and Lonergan.

[20] Cf, Bergson in *Creative Evolution* concludes (p.298) that "*a self sufficient reality is not necessarily a reality foreign to duration.*"

Is "being" more than this, in fact? Could it have an infinite potential? I believe that is something we will never know by examining our concepts alone. Is the "givenness" of being, its unquestioned solidity, its invulnerable *presence* — that feature that causes Heidegger to define being as "emergent power" — somehow possibly a **necessity,** an absolutely fixed timeless eternal infinity that stands in some kind of shared, creative, participatory relationship to this endlessly changing finiteness that we actually experience? What we are saying does not preclude that possibility, or others. What I am denying is that there is any way *to know it.* We cannot determine it to be *a fact* by simply analyzing our concepts, ... and our *experience* so far does not support such a conclusion. I am also saying that to claim that we can and do *know* it in this way is a projection of the human conceptualizing process. Rahner is right in this sense: *it's what we do.* It's a function of our psychology, our epistemology (our mystification of knowledge), and the less-than-complete state of our scientific understanding of the reality around us. I also believe it corresponds to a peculiarly human *paranoia*: we question existence at all only because we are in anguish that our death may mean our not *being-here.* On this, as in so many things, we are driven by our **conatus**.

the "metaphysical" question

This appreciation of the qualities and character of *existence* as opposed to "being" helps us understand why the metaphysical question of western culture, "*why is there something rather than nothing?*" required an extraordinary experience for its generation and justification. It was a shock to the **conatus.** *For the question is not normal.* The ordinary business of living operates on an unreflective background awareness of *presence.* Our organic apparatus takes existence for granted; we are endlessly oriented toward the future. If left alone, we would probably make no further enquiry into it. An indication of its *utter abnormality* is the fact that seriously questioning the solidity of *existence* immobilizes us. It seems to render empty and meaningless everything that we are or do. I believe this is a clue to the metaphysical issues that underlie our *being-here.*

Death paralyzes us. Projections of continued existence *in another world* are metaphors for the endless presence which is the natural orientation of our biological inheritance, *the* **conatus**. But we've just seen that what we call our demise is, in the worst-case scenario, simply a change in modality. In this context, then, our anguish about death reveals itself to be reducible to our attachment to our temporary experiencing — our "selves" — which we have, without justification, projected into an eternal entity and equated with *existence* itself. Once the question of the "disappearing self" is eliminated (and I mean as it has functioned for metaphysics, not psychologically), we realize that we, in fact, do not worry about *existence* at all. There isn't any "*nothing*" to compete with being. "Nothing" is not an option or a possibility. As far as the eye can see, in time and space, there is only being — wall to wall being — in a process of creative self-elaboration, change and re-cycling. All preliminary enquiry must arrive here and begin here. There is no real question prior to the questions, *what does being do?... what form does it take? ... why does it change appearance? ... what is it for?* In other words, we cannot ask, "*why there is something rather than nothing.*"

(What *does existence do*, by the way? It will be the principal corollary of these reflections about *existence* to claim that, true to its lack of "essence," it *has no purpose; it is* "*pointless.*" What it does is to go on *existing — it survives.*)

6. death and the primacy of "essence"

The "insignificance" of being, however, does not invalidate the personal and social importance of the *realization of the impermanence of the experiencing self* that we claimed was the source of all this questioning. The experience of death or its equivalent that gave rise to the anguish behind the "metaphysical question" has also served as the basis for profound personal transformations, considered important in the programs of the great religions, West and East. I'm referring to the sudden awareness that our "person," our "experiencing self" is not an entity that will endure. It is a shock. What occurs in this dramatic moment is the collapse of the illusion of an unending *experience-as-self.* The solidity and the assumed transcendent importance of the individual *ego* is fatally challenged.

But this disillusionment has been interpreted in two different ways. Hindu-Buddhism in its various forms believed that the "self" ceases as a center of experience but *existence* goes on intact in the form of the substrate of which our bodies and all things are constructed. Reality, in other words, belongs to the totality not to the individual. The Western tradition, on the other hand, invested as it is in the importance of the individual, was persuaded that it had to be the "vanishing" of the self that was the illusion.

It could not be that the individual person ceases to exist. Therefore the individual must have an immaterial spirit that cannot die with the body.

There is a paradox in Western thinking on this subject, however. Classic "being" came through and with the form, the *essence*. *Existence*, in other words, is secondary and subordinate to *essence;* it was *essence* that was the "carrier" of being. Matter, in contrast to form, was defined as *pure potentiality,* that is, in effect, "non-being." By defining death as the abandonment of formless "matter" by being-bearing "form," *existence* had to disappear because there was no *essence* to support it. So, Greek philosophy explained death as a return to nothingness. We have to realize that "immortality" in Greek thought is imputed to the human "essence" alone — the "form" or soul. It was proposed as *an exception to the general rule* about death. All other "forms" were material and mortal. When the "essences" left or were changed, the "things" died and disappeared from existence. The human form (soul) alone was spirit and could not die. The double standard apparent here is a clue to the untenability of the Greek conception. I believe the "immortality of the soul" is a metaphysical anomaly in the Greek system, explainable only as a course correction designed to offset the psychological unacceptability of human death. The **conatus** could not accept it.

This Greek theory of death was built on the control of "being" by form or essence. Without an "essence," there was no "existence." Once you stopped being "something" you stopped being.

In the East, on the other hand, the experiencing self is not considered an existing essence. So the shock of death leads to a different result: the awareness of *anatman* — that, despite the **conatus,** the permanent self is an *illusion* — which they dare to call *enlightenment*.[21]

The next chapter examines in more detail the peculiarly western way of looking at *existence* as a function of *essence*.

---

[21] The two Hindu "doctrines" are correlatives. Re-incarnation itself presumes that the re-use of the substrate reflects the lack of reality of the "self." The doctrine of *anatman*, then, is astutely used by Buddhism as leverage to project an end to *samsara*. Buddhism is simply one of the logical implications of *samsara-anatman* and the re-cycling of "being," which is the fundamental image dominating the whole Indian philosophical tradition.

chapter 2
the one and the many
essence & existence; participation in being

This chapter will attempt to draw out the initial implications of what has been learned so far. In the first chapter we tried to identify the basic problem of ancient philosophy by looking specifically at the foundational concept of "being" which that system produced and assumed as axiomatic. The traditional concept of "being," by projecting *intelligibility* as a primary property, postulated that invisible "essences" bearing purpose must be embedded in the "things" that exist. "Being" was therefore considered activated proportionately by the various existents according to the emergent level of their particular essence. The notion that ruled "essence" was *idea,* and the notion that ruled idea was *purpose.* Reality was intrinsically *rational* because there were *reasons* why things were what they were.

In contrast, we saw that our experience of the sheer *presence* of things irrespective of their level of emergence is totally homogeneous and indistinguishable in all things. It is not proportionate at all. We called it *existence* in order to distinguish it from the traditional concept of "being." *Existence* is univocal; it does not admit of proportionate predication. It is referring to only one thing ... the same *presence* we experience common to all things that exist.

In this chapter we will see that science confirms that *existence* is not proportionate because there are no "essences." What exists are various configurations of the same *material energy.* We therefore identify our homogeneous experience of *presence* with the homogeneous constituent of all things, *material energy.* *Existence* is intrinsic to matter. It is a self-embrace, the same for everything that exists. It is not derived from *an essence* introjected from another world — an outside *idea* bearing a purpose beyond itself.

There are also corollaries of "being" that need to be adjusted. The first will be to understand why the classic categories of "essence and existence which were the key components of traditional proportionate "being," functioned so well for so long, and why they are now totally *obsolete*. Later, the classic question of the "one and the many" will be introduced and it will be re-assessed from the perspective of modern science.   Another is the "principle" of *ex nihilo, nihil fit,* also the existence of "spirit," the prime corollary of the theory of "essences."   Then, how "participation-in-being" functioned in the world of traditional philosophy will be re-evaluated in a dialog with science.   Finally the implications of all the foregoing will be tapped for the construction of a new philosophical methodology and discipline — cosmo-ontology— to replace traditional metaphysics.   That is the agenda for this chapter.

1. essence and existence

Heidegger tells us that at the base of the ancient Greek view of the world there was a simple everyday image which has long since been forgotten — the potter and his clay.[22]   The potter takes a lump of formless common clay, throws it on a spinning wheel and, following ideas he has in his mind, makes items of various shapes, sizes and uses — cups, plates, pitchers, vases, trophies — each for a different *purpose* and therefore with a different *form*.

The common clay was the material, the "matter," that the potter's *idea* worked on.   The "form" given to the artifact copied a plan, an "idea" with a purpose, that originated in the mind of the potter.   It was the *idea* that was responsible for making a specific thing for a specific *purpose* out of general material.   With the image of the potter, the Greeks took what they knew of their everyday life and used it to try to understand the living species and the forces of nature they saw in the world around them, each of which appeared also to have a distinct form and discernible purpose.   On that basis they conceived of all reality as composed of *matter and form*; and the form was the expression of *an idea*, like the plan in the mind of the potter.   The *form* was the source of any given thing's "being" and purpose: *what* it *was* and what it was *for*.   Matter and form is the binary root of western thought ... and *form* was the dominant notion.

---

[22] Heidegger, *Phenomenology...*, p.106ff; cf also Eduard Zeller, *Outline for a History of Greek Philosophy*, Meridian Books, N.Y., 1883, tr. Palmer, 13th ed. Revised, Nestle, 1928, reprinted 1960, p.164 cites Plato's *Timaeus* ch.18 and 19 for the "craftsman" image in explaining "matter."

They called the *form* the "essence." Its *whatness* was its "essence" and its essence was its *purpose. What* it was, in other words, was what it was *for.* Matter, until essence made it "something," was "nothing." "Being," for the Greeks, came through the *form* for its origin and ground was the *essence* and its purpose.[23]

The potter imagery was "cut to fit" mediaeval Christianity, continues Heidegger,[24] because for the scholastics the universe was the work of a "Divine Potter" who brought things into being by giving them each a purpose embedded in an *"essence"* that was generated in imitation of the Divine Essence. Existence had to come *after* the essence. The essence was the plan, the purpose. It *had* to come first. And since the plan was for a "thing," existence came only as a "thing." It made dynamic *becoming* ancillary to static *being,* and put *existence* in service to "essence."

The long history in the West of the assumed unity of *essence and existence* — that actual existence is found only in conjunction with and *subordinate to* essence — comes from this primacy in Greek thought of the "idea," as the design from which things came into being. These terms, *essence and existence,* may seem foreign to us, but in fact they are responsible for and reflect the way we look at the world. The world for most of us, every bit as much as for the Greeks or mediaeval scholastics, is made up of clearly defined "things," each of which has its purpose, planned for it by its Intelligent Designer.

Essence is the counterpart in the creature of the *idea* pre-existing in the "Master Mind" that created it, like the plan for a tumbler in the mind of the potter. Essence embodied *purpose.* For it was clear that if the clay was made in the form of a vessel, it was only because its purpose was to contain. The scholastics, following Aristotle, called purpose the *final cause,* and it was the *final cause* that made things to be *what* they were. Its *purpose* made this thing to be different from what some other thing was. From this we can understand why ancient and mediaeval philosophy asserted that *existence* — which means not *what* but *that* the thing is — was secondary to, and came "after," *essence.* Essence was the ruling

---

[23] The form-as-plan also constituted the *final cause* for Aristotle — the form (essence) was determined by the *purpose* for which the "thing" in question existed. And. in turn, *purpose* was inconceivable apart from a planning mind. Cf Etienne Gilson, *From Aristotle to Darwin* ... Notre Dame, 1984.

[24] *Phenomenology* ... Op.cit.., p.118

notion. "Being comes through the form," was the favorite reminder of the scholastics. Actuality, *existence* was only a potential quality of the reality of the "thing." That the "thing" happened *actually to exist* or not was secondary. Bernard Lonergan in *Insight* quotes Thomas Aquinas as saying, "It is in and through essences that being has its existence." *To be* was to be *something.* Lonergan comments:

> "Hence being apart from essence is being apart from the possibility of existence; it is being that cannot exist; but what cannot exist is nothing, and so the notion of being apart from essence is the notion of nothing."[25]

We saw this in action in chapter 1. It reveals the primacy of *ideation,* concepts, intelligibility, for the appreciation of concrete reality in western metaphysical thought. It encourages *idealism,*[26] and ultimately makes "being" static rather than dynamic — a universe of fixed "things" — and all processes subordinate to their production.

being as process

All this has disappeared. Modern science has rendered the imagery of the potter and its associated metaphysical notions, obsolete. For *evolution* has revealed that phylogenesis (the creation of species) proceeds by adaptation to environmental changes "directed" by natural selection not by the "plans" of a Creator-Designer. "Essence" (a "plan" for a species) in other words, doesn't precede existence, rather *existence*, that is, **the drive to survive** precedes the form things take, not only for the observer, but for creation itself. The forms (which for the Greeks are *describable ideas*), don't come first; in the modern view, they come last. For "what" the thing is, is a function of its dynamic adaptation to environmental challenges ... it's drive to survive. *Existence* predominates. Species (essences) result *a posteriori,* forged by the creative dynamism of "survival," which is an *undefined process* whose end is its own undefined continuation. There is no "final cause" or ultimate purpose of the process of exist-

---

[25] *Insight,* pp. 371-2.

[26] Observe that western metaphysics as bequeathed to us by the Greeks, because of this subordination of existence to essence, is insuperably *idealist.* By that is meant that the reality of things is always secondary to and dependent upon the *idea* of things. The idea, and the Mind that produces it, is prior to reality in every sense except perhaps the chronological sequence of discovery. If it is further believed that the human mind has direct, exclusive and infallible access to these essences which are *ideas* (going beyond the limitations of time and space), it becomes entirely understandable that idealist *a priori* systems of thought would form a major if not predominant current in western philosophy.

ing. *Existence, being-here, has no purpose beyond itself. It is an inwardly focused self-embrace.* **The "purpose" of existence is to exist.** Hence "essence" no longer rules ... it follows and obeys *existence.*

The "principles of being" like *essence-and-existence,* once considered necessary to explain reality as a function of *ideas,* have disappeared. Darwinian natural selection as classically proposed means that the "form" that a species exhibits is purely functional, not teleological. *There is no pre-ordained purpose.* The "purpose" of the "form" is to serve the way the organism has managed to survive, it does not precede and determine the way it will survive. It's purpose (which defines its *essence*), is *existence.*[27] The form is the result of adaptation, not the cause of it. If we were talking to mediaeval scholastics, we'd have to turn their admonition on its head and say, "being does *not* come through the form, rather *form comes through existence."*

Evolution thus severs *existence* from its perennial subordination to *essence,* plan: idea and purpose. *Existence* without essence can be understood as the **dynamic self-embrace of existence itself** without an *a priori idea.* Without *essence,* "being" is revealed as *becoming*: a *process* extruded from a dynamism — an inwardly focused, self-reflexive, self-embracing energy that is self-explaining and self-sustaining. This *existence-becoming-existence* has no purpose beyond itself; it has no ulterior "reason" for being ... *therefore it can be said to be* **non-rational**. *Existence is a "mindless" paroxysmal dynamism* — **a non-rational process whose end is itself.** **Existence has to exist.** Following Spinoza I see this "necessity" revealed in a **conatus sese conservandi,** the drive for self-preservation that we experience within ourselves. We *have to exist!*

The ancient dualist-idealist metaphysics built on *essence-existence* as the necessary emanation of a Master Mind vanished when ancient science vanished. The new world view is based not on *two-principles of being,* but on *only one: the self-embrace of existence* which makes things evolve and we experience as the **conatus**.

---

[27] I am fully aware that the scholastics (and later Spinoza) said the only "being" whose essence it was to exist was "God."

2. *existence, matter's energy:* "one thing"

Our common-sense view of the world, suggested by modern science and our direct experience of universal *presence*, differs greatly from that of the Greeks. *Form or essence* for the ancients, as we saw, was the central concept that ruled the dyad of *matter and form.* It was the source of independent metaphysical "distinction." It *defined* one thing from another because it was the factor that made one thing different from another. We, on the other hand, take "form" in an entirely different sense. We see "form" as phenomenal, a feature of *function* not of "essence;" therefore form is secondary, possibly illusory and, at any rate provisional and temporary, for it will change when the survival needs change. Form, for us, does not *define,* and therefore it creates only *functional, not metaphysical distinctions.* Function is the result of adaptation-to-environment, *not* the result of an idea, the plan of a Creator's "mind." Form is no longer metaphysically relevant because fixed essences no longer exist. Nothing is fixed, everything is changing.

What provides the solid base for us is the **one, single, common and universal substrate** of the things that are — *material energy.* It is *what* all things are made of and what determines what all things do. *Matter's energy* is neither created nor destroyed. *It must exist.* This *insistence on existence* displays itself in the more developed "living" levels of emergence as a *drive to survive.* It is responsible for every feature and function in the cosmos, including evolutionary development. It makes all things "one thing" *metaphysically,* for there is no other "principle" by which things exist and do what they do. The distinctions between and among things are simply the functional re-arrangement of the elements of the same single substrate, a different intra-relationship of one physical-metaphysical "substance."

We know that *material energy* — in the form of particles, energy, radiation, light — is present in every entity that we know. In fact, given the almost immeasurable diversity of life forms, celestial and geologic forms, chemical and molecular forms, there is one thing that all these existent entities and forces have in common: *matter-in-process, material energy* in different configurations insisting on existing. Whatever distinctions these provisional forms display are merely the results of an internal redeployment of constituent elements. The temporary configurations elaborated by *matter's energy,* are a communitarian phenomenon. The unity of all things is grounded *intrinsically* in their constitutive possession of the same substrate ... and the substrate, *material energy,* as far as science can tell, is homogeneous and it is all there is.

Science has increasingly assumed that all reality was exclusively comprised of matter. By the late 19th century reality had been "reduced" in the common opinion, to a mechanistic determinism ... an alleged mechanical reflex. Those who battled it, like Henri Bergson, insisted that evolutionary phenomena could only be explained by the primordial presence of *spirit* guiding and enlivening the existence and forms of inert matter. This dynamism for development Bergson called *elan vital,* a "vital impulse." He felt forced to say the *elan vital* was "spirit" and "opposed" to matter because he was an idealist and a Cartesian dualist, and, like the reductionists,[28] also believed matter was inert. Please observe: the two notions are correlative; once you posit the dead inertness of matter you have no choice but to explain not only the vitality that we see all around us *but also existence itself* as a function of an outside source, a second principle ... or you leave "life" and *existence* without explanation. And since it's not "matter" what else can it be? Immateriality by any other name is still "spirit." This study will attempt to show that, to the contrary, *material energy* is itself, *a living existential dynamism.*

A corner was turned at the beginning of the 20th century when we discovered that **matter is, in fact, energy.** It's been more than a hundred years since Einstein worked out the equation $E=mc^2$. The math may be beyond us, but what it says is clear enough: energy and matter are convertible. Since the publication of that thesis in 1905, physics has achieved the inferential mapping of the interior of the atom revealing the existence of a myriad of "sub-atomic particles" all measurable as energy. This energy is on dramatic display every day in the conversions taking place in our sun where hydrogen fuses with itself to become helium and releases so much energy, at such a steady rate, and for such a long time that it has called forth and sustained every life form on earth for almost 5 thousand millions of years, and is not even halfway through its fuel stores.

Those protons, neutrons and electrons — the constituents of atoms — are themselves further resolvable into even smaller building blocks with

---

[28] "Reductionism" will come up frequently. It refers to a philosophical position sometimes called "positivism," that denies the existence of any reality not observed and measured by science in its current state.

strange names like quarks, gluons and neutrinos. The presence of these mathematically identified point-like particles (which also act like energy fields) is directly observable during the nano-second when, in their unattached, uncombined state, they convert from "matter" into energy leaving only a luminescent trail that disappears into the darkness of the cloud chamber. ***Energy and matter are different appearances of the same thing.*** This is a second factor that has upended the settled categories we received from our tradition.

The Greeks had no idea. And until the 20th century, frankly, neither did we. All of this tends to confirm our conclusions about the faulty dualist "concept of being" analyzed in chapter 1 and our decision to replace it with the simple, univocal *existence* — our experience of *homogeneous presence*. We experience that all things are *present*, because they are all made of *matter's energy*.

The most recent development in this area is an hypothesis known as string theory. By proposing that sub-atomic particles (or their "force fields") like quarks and gluons, once believed to be dimensionless "points," may be *vibrating strands of energy* with a very small but measurable extension, certain mathematical anomalies associated with the earlier "massless" view are resolved. At the same time the theory holds out the prospect that all of reality might be understood as a function of these vibrating string-fields, thus providing a model for understanding the presence of *energy* at the base of the pyramid of matter.[29]

But matter's energy is very different from the imaginary notion of "spirit;" *for it is not an idea* — the exclusive product of "mind" — a kind of reality allegedly opposed to matter. So we are justified in concluding: *material energy* and *existence are one and the same thing.*

3. the "one and the many"

Aristotle's rejection of Plato took over traditional philosophy definitively in the late middle ages. Aristotle differed from his teacher by saying that there were no independently subsisting *ideas*. There was only one thing that existed and that was the "substance," the individual existing "thing" made up of *matter and form*. The modern scientific world view, progressively taking root from the early 14th century with Duns Sco-

---

[29] Brian Greene, *The Elegant Universe*, NY Vintage Books, 1999. *passim.*

tus and William of Ockham has been focused on Aristotle's assumed exclusive metaphysical individuality of the "substance."[30]

But there's a paradox involved here.

Individual and species are correlative notions. The one implies the other. In fact the individual, by reproducing itself, making copies, creates the species. As Aristotle saw it, the individual thing (the "substance") is the more fundamental. It is the *locus* of being. The *species*, he said, is not real. A group of individuals may exist, but a *species* does not; it is only a mental construct. Individuals and their species are the original bases of the classic "problem" of the "one and the many."

Let's take a minute to understand just why the "one and the many" was a conundrum for the Greeks. Consider: you see individual entities, let's say ducks, walking around ... each one obviously independent of the other. If you're interested in "being," as Aristotle was, you'll say, that duck has it's own "being" because if all the other ducks were to die, it would still be there. But how, then, can an individual duck *be exactly like all other ducks.* That which makes this duck a duck, is not just hers, it belongs to them all. The "duckness" is the very same ... does that make them all one duck? How can they all possibly be the same ... and at the same time be so different as to exist independently of one another? Isn't the one living duck, for being the same as the others, somehow not independent?

The theory of "essences" solved the problem of the one and the many by identifying where the "sameness" resided — with the *one* essence (*form*) of "duck" ultimately lodged in the "Mind of God," providing existence to multiple specimens through individuated matter.[31] But once evolution demolished the Greek theory of matter and form by showing that *form* (essence) is not a guiding plan, but a more-or-less arbitrary byproduct of survival, that theory no longer explained the relationship between the individual and the species.

Well, then, what *does* explain how there can be both "one" and "many" of the same thing? If, as we are saying here, there are no longer any

[30] Please note this nomenclature. "Substance" for traditional scholastic philosophy and its inheritors means "thing," not a chemical or mineral.

[31] There is a problem with hylomorphism (the theory of *matter and form*) as an explanation for individuality, in that the "matter" envisioned was an amorphous "prime matter" which received all its determinateness from the *form*. If the *form* is generic (e.g., "duckness") and not individual, where does the individuality that separates one duck from another come from? *Prime matter* does not have the wherewithal to do that.

forms, what makes a duck different from a horse or a human? And without *matter accompanying form,* making it an individual apart from the rest of the species, how can this be an *individual* duck?

Stephen Jay Gould: what is an individual?

I would like to illustrate the question of the one and the many by tracking the way *individuality* was discussed in a book on evolution by the late Stephen Jay Gould called *The Structure of Evolutionary Theory.*[32]

We join Gould on page 602 halfway through his enormous tome and elaborate argument. At this point he is looking for the *locus* of natural selection, i.e., *what* was being selected. Was it the individual organism, he asked, or was it the gene, or perhaps the species?

Before starting down this road let's take a minute to remind ourselves what "natural selection" means. It was a metaphorical term chosen by Darwin himself inspired by the methods used by people in the breeding of domestic animals. People knew how to "select" among various individual animals and breed them so that certain desired characteristics would appear and eventually dominate all the offspring. This was the image that Darwin used to explain how *nature* "selected" among the various organisms — granting "reproductive success" to those individuals who displayed desired characteristics and eventually new species. It seemed beyond dispute that the focus of "selection" had to be *the individual* who possessed the desired traits.

So Gould set out to determine what makes "an individual," for only an individual he thought, following Darwin, could be the target of natural selection. Now let's consider. A human arm has a certain amount of individuality, but you wouldn't call an arm "an individual" and it is not the instrument of evolutionary change. In order to distinguish things that only *look like* individuals, from those that really are, certain "vernacular" criteria were initially proposed by Gould to identify individuality. He suggested *birth, death and constancy of form* (leading to continuous recognition). Under those criteria, however, he found there were a vast number of identifiable "individuals" — even the earth itself — which in fact *lived, died and remained recognizable* but did not carry forward evolutionary change. Were they individuals in the Darwinian sense? He couldn't answer the question about individuality, but then, suddenly, he realized *he didn't have*

---

[32] Stephen Jay Gould, *The Structure of Evolutionary Theory,* Harvard U.Pr, 2002, pp 602 – 613

to. He could as easily work backwards. For as far as his search was concerned, the individual was whatever "entity" happened to be the bearer of evolutionary change, i.e., the carrier of the desired trait. Instead of identifying the individual and then determining the function, he let the function determine the individual.

So he decided to identify the bearer of evolutionary change, and then call it the "evolutionary individual," without worrying about whatever other individuality may exist at another level. For this, Gould proposed criteria that were specific to the requirements of Darwinism: *reproduction, inheritance, variation and interaction with the environment*. By these standards it became clear that many "entities" that weren't individual organisms, like a bee hive or a fungal spore colony, could still be considered "individuals" from the point of view of "natural selection" because they carried forward evolutionary change under the four criteria.

Without directing themselves specifically to the philosophical question, scientists like Gould realized that their interest in the question "what is an individual" had to do *only* with the subject of change and therefore it had no other basis for its resolution than **function**. In the two cases just mentioned, the bees and the mushrooms, the collectivity bore the burden of the function. That was all the scientists needed to know. *They didn't care about any other kind of individuality,* and given the absence of any acceptable "metaphysical principle" of individuality, as far as Gould was concerned, there *was* no other kind.

Determining individuality would seem to be as simple as saying, "if it looks like a duck, walks like a duck, etc., then it *is* a duck." But scientists didn't care about *what* the thing "*really* is," because for them **its reality is its function**. Gould's only concern was whether the "entity" carried phylogenetic modifications into the next generation. He was content to say, "this thing, whatever it is (even if it is an appendage of something collective, or, as in the case of the bee hive and the spore colony, aggregates of smaller organismal 'units'), is what carries evolution forward." On the basis of these latter criteria, then, Gould justified calling the *species* an "individual." This is what he was really after. It supported his thesis that evolution *also* operates at levels higher than that of individual organisms. He considered this an important element in his theory of "punctuated equilibrium."[33]

Gould's theory recognizes that *stability* ("equilibrium") is one of the marks of successful species and it explains why the evolution of species is "punctuated" and not uniformly gradual. Species, in other words, once they've achieved a viable adaptation, tend *not* to change. There are identifiable biological mechanisms, "constraints," developed by species to prevent or at least minimize genetic drift and the emergence of mutations. (Why these "constraints" seem to lift at a certain point and allow for a profusion of new speciation "punctuating" an ocean of stability is a reality that Gould's theory sets out to explain.) Such stability, however, must itself be recognized as *functional.* A species is not a fixed "essence," but rather a *temporary stasis* as a survival strategy within a changing environment.

All species are in constant flux, even though it is not observable on our time scale. 99.9% of all species have gone extinct, which only means, of course, that they have changed, i.e., they have evolved into other species, and of those that have not, none have gone unmodified. Even today's bacteria are not simply the clones of their ancient ancestors. The stability behind "punctuated equilibrium" is relative. It is, like any individual, temporary and provisional. If the form and function that an individual possesses at any given moment is so temporary and provisional, in what way can we attribute "reality" to it?

The thing to notice is that once you claim that there is a difference between the two "kinds" of individuality — appearance (function) and "being" — in other words, once you seriously ask *what things really are* in spite of what they might *look or act like,* you have left scientific discourse and you are asking a question whose only answer can be an "essence" or some equivalent. Your question itself assumes the static "essentialist" character of reality and ignores the possibility that "being" may be a homogeneous substrate (totality) in process and that what appear to be individual entities are secondary to it.

Scientists, on the other hand, *cannot* ask the question about what things *really are, because they have no interest whatsoever in the answer.* This is not a personal choice on their part; it is the nature of "science" as we have inherited it.

---

[33] For a thorough discussion of "multi-level selection" cf Bert Hölldobler and Edward Wilson, *The Superorganism,* NY W.W.Norton, 2009, chapter 2.

But if you were to say that the question *cannot* be asked from any other point of view than the strictly functional and phenomenal, as science does, then please notice, it means you are *ignoring the traditional philosophical basis for individuality without substituting another*. If identifying "being" at a level below the totality of *existence* (the entirety of *matter's energy*), can only be tied to its function, then ontologically speaking, there is no way to deny the claim that there are no "individuals" except in appearance only ... but neither does it prove it. The issue, in any case, is of no direct concern to the scientist. Whether the subject is an ontological individual or a mere appendage of some dynamic aggregate (or of the totality-as-energy) is irrelevant to the scientist who simply wants to explicate the function or structure in question. Individuality, beyond its role in sustaining a function, is irrelevant.

We conclude: the assertion of metaphysical individuality is a derivative of essentialism. Without *matter and form* (or some other similar ontological basis) *there is only functional individuality*. That means that, metaphysically speaking, we cannot eliminate the possibility that there is only "one thing" in existence: this universe — one single organism, with an *existential* dynamism that unfolds into everything that is. Spinoza identified "everything" as one "substance" ... was he right?

What gives us the right to say that any given phenomenon exists as anything other than a modality of some "one thing?"[34] Some may answer, "death." When one individual dies another is unaffected; doesn't that establish individuality? But if one cell dies, other cells and the body itself is unaffected. Organisms live and die and the species remains unaffected. What's the difference? We've already seen in chapter 1 that death as a proof of existential extinction is deceptive. With Gould we realize "death" is not a criterion for individuality. What makes the apparent individual "thing" anything more than a cell of a larger organism, or an aggregate of smaller organisms? What if "the race," or "the nation," or even "humanity" is the individual? Wouldn't that make individual persons only "cells" of a *superorganism* like the cells of any living thing? We will recognize that in

---

[34] Baruch de Spinoza (1632-1677) held that there is only one *substance*, God, whom he equated with Nature ... all other "things" were only *modes or modalities* of God

the history of western philosophy, religion *and politics,* these are not entirely unthinkable questions.[35]

From this analysis we can see that there is an ***irremediable incompa-
tibility*** between these two world views as we have inherited them — the
traditionally philosophical and the traditionally "modern" scientific. To accept the reductionist scientistic world-view, unmodified, as if it were a metaphysical fact, is to eliminate the basis of intelligibility for the Greek ontological world-view, and vice-versa. Scientists, were they to think about it
"metaphysically," would be logically incapable of identifying "being" with
anything but the totality-in-process *because they reduce everything to the
substrate, material energy.* Ontological individuality below the level of the
undefined dynamism of the substrate (i.e., the totality) cannot be assumed, may be unprovable and, since it is irrelevant, tends to be ignored.

six levels of individuality

So Gould concludes that functional individuality for the purposes of
evolutionary science, exists (potentially) at *six* different levels: *the gene,
the cell, the organism, the deme* (an isolated sub-set of a population), *the
species and the clade,*[36] because all can carry evolutionary change.[37]
What "is-ness" might be besides this, he is unashamed to say, is not his
business.

Classical essentialist philosophy explained individuality as a function of
matter and form. The *form* corresponded to species, made a thing "what"
it was and sharply distinguished it from the things of other species. *Individuated matter,* then, was supposed to separate the members within the
species. If we eliminate essentialism, it means that the locus of "being" is
no longer spontaneously identified with the individual organism grounded
in the existence of *forms or essences.* And to identify "reality" with the
totality (the substrate) gives us no new information at all. It leaves us exactly where we started: with the experience of finite, limited, *matter's
energy,* undefined except by its temporary appearance and provisional
function, integrated into apparent "units"[38] at various levels. It is precisely

---

[35] The allusion here is to the personification of collectivities like "The French Nation," "The White
Race," or "The Mystical Body of Christ."
[36] The "clade" is the evolutionary biologist's equivalent of a "class." Where they differ is in the
criteria used to determine the affinity among species. For the "class" it is perceptible similarity; for
the "clade" it's common genetic parentage regardless of appearance.
[37] Gould, *op cit.,* p. 681, *passim*

where the elimination of "essences" could be predicted to leave us —
back down to a groping, roiling universe, hopelessly indefinable, without
an invisible celestial scaffolding, interpreting what's unclear in this world
by imagining a fixed structure in another world. It is the recognition of the
irreducible primacy of the unprogrammed becoming which is existence. It
is the re-installation of the central paradox of "the one and the many,"
which the Greek philosophy of "essence and existence" was supposed to
have resolved.

Once Roman Imperial Christianity took control of Greek philosophy,
the *amorphism* that lurked behind the question of the "one and the many"
was not only a paradox, it became a *problem*. The authorities were determined to locate a metaphysical basis for what they were convinced was
fixed, stable, "being" in the *individual human person,* whose moral com-
pliance was tied to an eternal *individual* reward or punishment after death.
It was essential to "crime–prevention" and submission to the authorities.
Moral imperatives could hardly be imposed if the person were not an ontological individual, with a separate, unique and eternal individual destiny.
This moral emphasis on individuality, characteristic of the "christian" west,
would eventually lead to an exaggerated individualism that characterizes
European culture and the lands it colonized. *Matter and form* provided the
explanation they were looking for. Unfortunately they had to go to an
another world to find it.

The new perspective we are assuming here, on the other hand, allows
us to live in the same world that we see in front of us. For once we abandon ancient naïve assumptions about *metaphysical* individuality and its
necessary ground in essences, there is no "problem." It does not matter
that the person may not be a "stand-alone" metaphysical individual. Moral
behavior responds to other — communitarian — priorities! There may be
more *group responsibility* for individual behavior than our individualist inheritance has been willing to entertain. And there is much in the vision of
our first century christian witnesses — generally ignored — that speaks of
a cosmic, i.e., collective *eschaton.* But more of this later.

---

[38] David Bohm, in his *Wholeness and the Implicate Order,* calls them "vortices," like little whirlpools
that coalesce within a flowing fluid. NY Routledge, 1980, 2002, p.12

4. *"ex nihilo, nihil fit"* — an anomaly in a process world.

The capacity for consciousness in later sub-species of *homo* were not determined nor predicted by the capacities of the hominid species from which those changes emerged. Phenotypal differences are a matter of genetic adaptation between organism and environment; they *end with an achievement that they did not possess at the start.*

This will demand another important modification in our traditional "metaphysical" principles and apparatus of inference. We can no longer say *ex nihilo nihil fit*, "from nothing, nothing comes," meaning that the change had to have been *already contained* in the organism that later changed. As traditionally understood, that was an overstatement. The *potential* for that particular change was there, of course, and verifiable because of the observed emergence. But we have no way of knowing what "potential" means in any given case. What things are now does not define what they *were*, much less what they may eventually *become, and it does not identify **all the agents of change***. Change is produced by interactions involving the whole *totality-in-process* — the whole environment, proximate and remote — and affects the "substance" *intrinsically*, not accidentally. "*Ex nihilo nihil fit*" cannot be taken in its traditional meaning; *for its traditional meaning is limited to what is implied by fixed essences* supporting metaphysically individual "substances." In a world without essences anchoring a multitude of metaphysical individuals, it should not surprise us to find that "metaphysical principles" like *ex nihilo nihil fit* apply only to the one "substance" there is: *matter's energy* as a totality.

Potential requires activation. That means the activator(s) (whatever they might be) are as integral to the process as the potential. Why the potential / activator relationship should go so regularly unnoticed or misinterpreted in philosophical discourse is an indication of the blinding power of the theory of fixed essences. It includes the whole environment and is the dynamic operating in all organic change.

Applying the maxim **only to the organism** *a quo* has gotten us into trouble before. For example, mediaeval theologians were convinced that the "spiritual" capacities of the human being were *definitely not* contained in the purely "material" sperm and ovum from which the human organism developed. The axiom *ex nihilo nihil fit* drove the schoolmen to conclude that therefore "God," miraculously and outside the processes of nature, must infuse a "soul" into each developing fetus.[39] But we know that the human sperm and ovum actually *become* a human being after a process of cell division, multiplication and specialization under the precision guidance of the combined parental DNA leading to the perfect development of an intelligent human being. In mediaeval terms, *the germ cells become something they were not* and therefore contradict the maxim, *ex nihilo, nihil fit.* Similarly, the bovine sperm and egg become bulls and cows, and the poultry sperm and egg become roosters and chickens. No one of these developments is any more miraculous or "transcendent" than the other, requiring the direct creative intervention of "God." In each case *something new comes to be from something that it was not.*

That is the nature of organic change and it reveals the "nature" of *real potential.* One-plus-one, in the real world, as a matter of universally observed fact, does not equal two — it equals three or maybe even more. How it is to be explained may remain problematic, but its observation as an *undeniable fact* is not. Potential is a capacity of *material energy*, a power (*potentia*) that resides in the substrate. That fact gives us no specific information at all. For to say *ex nihilo nihil fit,* meaning that "the effect was **somehow** contained in the cause(s) *as a potential,"* is simply to restate our raw observations in other words. *Matter's energy* produces new being *a posteriori* from a potential that it had *a priori.* By knowing *that,* **we know absolutely nothing** except that the power to bring forth new forms of *existence* is an intrinsic property of *matter's energy.*

Real potential must *really be activated,* but the activator(s) in all cases are also *matter's energy.* And we have no way of knowing "how much" is due to the "thing's" potential or the activator's stimulus. These processes are so commonplace and universal that we tend to miss their significance: *new things are really brought into being out of something that they were not, through the real agency of a material "creator" acting with material energy's real potential.*

What things become, in other words, is as much a product of the *process* as it is of the "thing" from which they originate; and the process includes input from the entire environment. So, we can say that the traditional application of the principle *ex nihilo nihil fit* to pertain to the *"substance"* (Aristotle's "thing") *alone* has been improperly narrow, and this on

---

[39] This is a position on which the Church continues to insist. Cf "The Catholic Catechism" Vatican Press, 1991, ## 365, 366.

two counts. The one is, what we just mentioned, that it does not consider the environmental conditions in which development occurs to be *metaphysically significant*. And the other is that the very substance itself, the "thing," the subject of change, is not a "fixed essence" as assumed by the traditional categories. The findings of science indicate that the "base from which" must be understood to be ***itself a process***. The change occurs in and from the substrate, *material energy,* which is both organism and environment. The demands of survival in the context of that dynamic holistic identity explain this universal need for an adaptative *process*.

Furthermore, adaptation is the source of variation. And without that variation, as Darwin said, natural selection has nothing to work with.[40] That is, its metaphysical "nature" must be seen as not just passively malleable, but *actively adaptive*. Change is not something that only happens to it. The organism itself is biologically involved in its own "substantial" modifications.[41] Gene mutation, DNA drift and the loosening of the constraints on mutation[42] are intrinsic to this process. This unveils a property of reality not readily perceptible on our time scale. The ability to change seems to be a fundamental feature of matter energy's embrace of *existence*. It suggests, as we have noted throughout this study, that entities we have traditionally considered metaphysically "separate" and independent fixed substances can in fact be better understood as the modalities of a single substrate, which suggests the metaphysical identity resides in *the totality* ... what Spinoza called the only "Substance."

The ancient metaphysics of fixed and eternal forms, "essences," supporting independently subsisting substances did not prepare us for this. Evolutionary change is not limited to (in Aristotle's terms) "accidental" modifications of "substances." It actually creates new "essences" i.e., new species. If we were to use traditional terms, therefore, what we have in "evolutionary process" are *metaphysical modifications*, elaborated incrementally, involving the participation of all elements — the subject of change together with the environmental conditions that act upon it — all contributing in their respective measures to the final "achievement," which

---

[40] Charles Darwin, *The Origin of Species*, Chapter "Natural Selection" reprinted in *The Making of Society*, ed. Robert Bierstedt, Modern Library, 1959, p 244. Darwin says, *"Unless* [profitable variations] *occur, natural selection can do nothing."*

[41] Cf the discussion of "divergence" in Chapter 3

[42] For the phenomenon of "constraints" to DNA drift and what it means: cf Mayr, *op cit* p.106 and 113 the bibliographical footnote #2, which includes citation to Gould.

is *essential* change — speciation. So *ex nihil,* surprisingly, **aliquid** *fit ...* out of nothing, **something new** comes to be.

5. spirit and matter

Of all the primitive, spontaneous and uncontested assumptions of the human family, the notion of "spirit" is possibly the most universal and perennial. It crosses culture and time barriers that have proven insurmountable for other less transcendent factors. It was a central element of the view of the world that we inherited. For the Greeks from Plato onward, spirit was a given.[43] And when they asked *what* it was, what it did and how it could be understood, the word they chose for it, "spirit," contained the seeds of its explanation.

"Spirit," Latin *spiritus,* accurately translates the Greek word, *pneuma,* whose original concrete meaning in both languages is "breath." Breath is the sign of life. Breathing stops when life has departed. This invisible breath, by which we lived was gone at death. The Greeks naturally concluded that *pneuma* was an invisible "something," separate from our bodies because at death the body, whole and entire, was still there even though its "breath" had departed. This "breath" takes our very "selves" with it when it goes.

On this basis then, in conjunction with the analysis of the "potter's" invisible, hence presumably also *immaterial idea* putting purpose, "form," into material clay, the Greeks constructed a world of imagined *invisible reality* that helped them "understand" what they already "knew" of the visible world. The world they saw was explained by a world they couldn't see, a world of ideas in the Mind of God. "Mind" was the key category. "God's Mind was behind it all, and our human minds were able to discern it all, through our ability to grasp ideas.

Spirit, *breath,* was life; and each kind of life had to be the *form of an idea* — an essence Without breath-essence, without the *spirit-idea-form* of human nature, even the most magnificent specimen of a human body in short order deteriorated into an unrecognizable mass of amorphous "matter" that converted back into earth and dust. The Greeks came to believe

---

[43] Zeller, op.cit., *Outlines ..*, p.47ff, says that the notion of the immortal human spirit was originally foreign to the Greeks. He identifies **Thracian Orphism**, which he claims is of eastern (Persian and Indian) origin, with being the source of the belief in the immortal soul in Greco-Roman thought. This doctrine was diffused by the Pythagoreans, and from there to Plato who was its primary proponent.

that the body, *flesh,* without the spirit-form is dead and meaningless. Matter, without spirit — body without mind — is "nothing."

From the identification of idea-form with spirit and life, the rest of the Greek conceptual framework fell into place: "God" as Master-Mind; Creation as "God's" self-reflective *thought*; the unique *thinking* human "spirit" made "in the image and likeness of 'God;'" the immaterial transcendence of mind and thought, individual spiritual immortality — and throughout, the inertness and "nothingness" of matter. This "philosophy" was elaborated with the consistency of a Euclidean theorem. It produced a mental geometry that traced and measured the architecture of an invisible parallel world, the sacred basis *and explanation* of our rapturously beautiful cosmos. It was an imaginary edifice as symmetrical and transhistorical as the granite temples of the Olympian gods.

The central proposition which the Greeks handed on to mediaeval philosophy was that *idea-form* was also *spirit-life* and stood in absolute irreconcilable opposition to dead, inert unthinking "matter." It is this classic Western foundational duality, reinforced and welded to science by Descartes in the 17th century, that is now under challenge.

chomsky's challenge

At the end of his "Managua Lectures" of 1988, [44] Noam Chomsky declares himself open to the possibility that "***matter***" may be more than simply the passive recipient of motion. The significance of that statement goes far beyond its place in his presentation. For Chomsky is suggesting that for the last three hundred and fifty years science has been performing its tasks without a definition of matter.

The historical background for this anomaly has to do with Descartes' definitive division of all reality into two "things," called *res extensa* and *res cogitans* ... literally, whatever exists, will be either "an extended thing" or "a thinking thing," — mindless matter or rational spirit. It is an early scientistic restatement of Plato's dualism. Today the terms used are body and mind.

It may be difficult for us to appreciate the radical change which Descartes' distinction represented for the way matter was understood in 1641. He replaced the Aristotelian-Scholastic conception of an amorphous *prime matter* by *redefining matter* as a separate "corporeal substance," which,

---

[44] Published as *Language and Problems of Knowledge,* Cambridge, MA, MIT Press, 1988, p.142ff.

regardless of form, "possesses properties all of which fall within the pur-view of mathematics."[45] What it meant for Western thought in the long run is that essentialism was replaced with reductionism and the existence stu-died by science became totally fibrotic — lifeless, dead — measurable but otherwise meaningless.

Matter for Descartes was a purely passive inert "substance" limited to the kind of interaction that Chomsky describes as "contact mechanics." Spinoza — who was thoroughly Cartesian in this regard — defined it clearly: matter can be acted upon but cannot act.[46] Matter's inertness and passivity was so thorough and undeniable in their view that, according to Descartes, it was able to serve as a reliable indicator of the necessary existence of an invisible "second substance," mind, needed to explain "that which went beyond the properties of matter."

Descartes' conception of matter was, ironically, refuted by the near-contemporaneous work of Isaac Newton who discovered that **gravity** was a force that "acted at a distance" *without physical contact of any kind.* Newton was himself a Cartesian and was perplexed by his own discovery, and despite the evidence that *the limitations of matter are, in fact, **not** known,* insisted that the explanation must be framed in mechanical or qu-asi-mechanical terms. What that means, says Chomsky, is that science has been operating without an adequate notion of the matter it studies. This should have undermined the theory of matter's definable passive properties and erased the claim to know the existence and activities of the "second substance," mind, which was predicated precisely on knowing the where the properties of "body" stopped.

The general conclusion for Chomsky is that the Cartesian concept of "body" is untenable. "In short," he says,

> there is no definite concept of body. Rather there is a material world, the properties of which are to be discovered, with no *a priori* demarcation of what will count as "body." The mind body problem, therefore, can not even be formulated. The prob-lem cannot be solved because there is no clear way to state it. Unless someone proposes a definite concept of body, we cannot ask if some phenomena exceed its bounds.[47]

[45] René Descartes, *Meditations, Med VI,* tr Lafleur, NY, Bobbs Merrill, 1960, p.134
[46] *Ethics,* II, p13 *passim*
[47] Chomsky, *op.cit.,* p.145

He doesn't attempt to solve the problem, being content with opening the door to the *material evolution of the language faculty* which he believes is a hard-wired feature of the human brain — a point of his lecture. But the larger problem is exactly what we are setting out to solve in this study. *If matter is not defined by "contact mechanics," just what is it?*

We, in our time, have learned, that *matter* is *energy*. Matter is the unique source of all the energy in our cosmos, including the energy of existence itself, the energy of life and its specifically human manifestations like mind and thought. There is no longer any need to "explain" life by something outside of, other than, and separate from matter, for matter is no longer "matter" as it was for the Greeks. Everything that spirit was once called upon to explain — vitality, autonomous activity, self-identity, perdurance, adaptation, consciousness — is now intelligible as the properties of *matter's energy*. Spirit and matter, for us, are one and the same thing. This doesn't reduce all things to "mere matter" any more than it defines all reality as "mere phenomena" of the Spirit. *It contemplates something far more radical: it obliterates both those categories altogether, for without a basis in fact, "vital spirit" and "inert matter" are seen to be equally imaginary projections of a scientific world view that has disappeared*. "Spirit" and "matter," equally, no longer refer to anything real. The most they can be used for is the identification of past ideas. They are museum pieces, historical oddities.

spirit and the sacred

We are confronted with an entirely new reality. I have chosen to call it *matter's energy*. It is reality as we know it. It's the way things are for us. It is *existence*. We're in a completely different world from the one the Greeks and their disciples tried to understand. Essence and existence, matter and form, mind and body do not explain this new world. The entire architectonic structure that was built on these foundational Greek dualities disappears as well, including the traditional "Sacred." The Greek dualist vision had equated the Sacred exclusively with "spirit," understood as a separate, immaterial genus of "being" that could only be known by a separate, *immaterial* cognitive apparatus — mind — exclusive to human beings. In our world, however, this genus of being called "spirit" does not exist, nor any unique *immaterial* human mental apparatus that could be said to know it. We know that our mental operations derive from and are totally limited to the capacities of our organic infrastructure, *our flesh*. And we know our bodies are the elaborations of the material substrate of the universe. Energy is now the source of life, and the role of "spirit" as conceived by Plato and Descartes, has disappeared — absorbed into the transcendence of *matter's* energy.

How do we deal with this disappearance? How do we understand it? What does it do to our sense of the Sacred?

6. participation

T he major derivative of the *essence-existence* dyad for traditional essentialism was a philosophical concept known as **"participation-in-being."** It was a further expression of how the many are one. It was also the source of the sacredness in things. Traditional Western thought of the Platonic / neo-Platonic school projected that all metaphysical relationships were grounded in "participation-in-being" ... and being was "God."

"Participation" meant that "things" are not discrete, separate, individual entities as they appear, but rather, through their essences, exist at more than one level simultaneously, the most real of which is shared with other things in a transcendental pool. Reality was collective, generic and believed fully observable *as ideas* only by immaterial mind. This collective immaterial dimension derived from the Platonic belief that the *essences* which defined and gave *reality* to things, were universal concepts that subsisted independently *as ideas* in another world — which the Christians identified as the "mind of God."

As Christianity first embraced and then in its Roman Imperial phase dominated the Platonic / neo-Platonic world view, it assimilated *participation* to its doctrinal base. Plato's World of Ideas became the Mind of "God." "God's" creative ideas, the essences, were considered the *truly real* realities because they were generated and resided in the Source of all Being, "God's Mind." The concrete "things" they sustained in this world were like shadows that existed only derivatively — by *participating* in the reality of the creative ideas in "God's" mind.

"Reason," rationality or *Logos,* for example, was considered more than just an idea or a faculty. It was a *real being* apart from other beings, ultimately an emanation of "God's mind," and therefore a divine entity, whom the Christians identified with Christ. Human beings were rational and so *"participated"* in the reality of *Logos* making them rational "spirits," to some degree divine and naturally immortal. It was not difficult for them to imagine that all of humanity was destined to be Christian because they were already "one" by nature, united through *participation* in the super-rationality of the *Logos.*

Needless to say, the condition of possibility for all this "participation" was the *immateriality* of the independently subsisting ideas, the *essences,* which permitted mutual compenetration and inclusion. Matter was excluded from this *participation* because it was considered dense, impenetrable and absolutely singular. It resisted compenetration. Participation was an exclusive feature of *ideas,* and spirit.

The very *"concept of being"* itself is the most important example of how *participation* functioned in the ancient system. Mediaeval theologians utilized the logical inclusivity existing between the concept ("being"), and the universe of actually existing entities ("beings") in order to root the intelligibility of the creature-Creator relationship in deep metaphysical soil. For them "being" was not only an *idea* but an actual cosmological reality called "God." The creature's being had to share the very being of "God," for, as we saw in chapter 1, *the concept* indicated there could be only one "being." To be therefore, was to share "God's" being. That "sharing in being" was metaphysical participation; it was what it meant *to be.* Traditional scholastic metaphysics was not only a "science" that explained how things could "be," it also limned a relationship that opened to a profound mysticism grounded in a radical communitarianism. Participation-in-being made a sacred collectivity of the whole universe *in "God."* All this could be found inchoately in the NT epistles of Paul, presaged even earlier in Philo of Alexandria.

It was also felt that all *ideas,* including the idea of "being," were the only "real" realities; they provided knowledge that was *superior* to the observation of actual entities in the world of experience, which latter were considered to be only imperfect shadows of the real. The fact that Socrates may have been born without a nose, or lost it fighting at Thermopylae was data that belied the general truth that "all human beings have noses" predicted by the universal essence. Individual variations deviated from the immaterial "ideal" and were therefore held to be "inferior" to the "perfect" knowledge provided by the contemplation of essences alone. According to that world-view, concrete fact, *precisely because of its association with matter* which was responsible for individuation and defect, could be safely disregarded in the search for truth. The concept of "being" was therefore believed to give us more reliable information about "God" than anything outside of "revelation."

class or clade? biology trumps logic

Modern science stands at the other end of the spectrum from these ancient notions based on conceptual classification. The participatory vision projected by traditional metaphysics began to evaporate when mediaeval thinkers like William of Ockham came to agree with Aristotle that Plato was wrong. They said there are no independently existing ideas; therefore *"essences,"* do not exist outside of the individual and cannot be the source of any collective reality. *"Participation"* is an illusion, they said. Individual entities are all that exist. The class to which something belongs is only a *mental construct*; it does not exist on its own and therefore it does not ground a participatory unity higher than the individual. For Aristotle each individual entity exists separately from every other, even those of the same species. The influence of one such entity on another was merely extrinsic; any relationship between them was effected by one acting "outside" itself (*ad extram*) on another like a billiard ball.[48]

Under Aristotle's tutelage, scientists began focusing on concrete individuals and *not* on the essences that defined them. The connections among entities were recalculated through the observation of actually existing reality. This resulted in modern science rejecting any claim that word-based "class" is an accurate guide to the biologically produced speciation evolutionists call cladogenesis or phylogenesis.[49] Science says the adaptive modifications of later groups of species, **clades** (the word has been recently coined by biologists, inspired by the word *klados,* Greek for "branch") are possible because the inheritor species utilize structural patterns and functional apparatus developed by their ancestors. This is evolution. There is an ever expanding branching development of species which can be recorded in tree-like genealogical charts called ***cladograms***. Each branched outgrowth is a biological daughter of its parent species.

---

[48] Substances for Aristotle were the instances of "essences" and hence, besides their dense individual materiality, were metaphysically distinct from other essences, even those of the "class" to which they belonged, and impenetrable. A being that was penetrated by another, metaphysically speaking, was *essentially* compromised; it was consumed or fatally modified. It stopped being what it was and became something else.

[49] Ernst Mayr, *Toward a New Philosophy of Biology*, Harvard U. Press, 1988, p.269

So according to the cladograms, the human organism can be said to have genetically inherited the biological accomplishments of a long line of pro- genitors reaching back well beyond the primates into the earliest mamma- lian vertebrates, who besides having brains, spinal chords, hearts and lungs, muscles and bones like the earlier reptilians, also gave live birth, suckled their young, were able to regulate heat internally (characteristics of mammals alone), all of which they passed on to us humans as well as to a myriad of other species of animals. We all enjoy these *same* marvel- ous features of our body-persons because we are all genetically related; we are their "clade," their progeny, their family, and we *participate* in their organic achievements. They provided us with the very structures that make us ourselves.

So, once these biological relationships are reliably established, it might not be entirely inappropriate to speak "scientifically" of the *participation* of species in one another's phylogenetic accomplishments. "Participation" in this case, however, is quite different from the ancient conception based on class alone contemplating a shared immaterial "essence;" but it's similar in the sense that the genetic configurations — making things be *what* they are — is shared. Phylogenetic inheritance as expressed in the clade de- emphasizes the separations between species (and of course, individuals) and reveals an intimate relatedness that shows the many to be one.

Of course, there is nothing particularly arcane in a relatedness of this type which is simply the extension of the familiar phenomenon of **repro- duction,** providing genetic continuity. Parents give of their own cellular reality under the guiding hand of their shared DNA to produce another in- dividual organism that is in every respect the biological outgrowth (includ- ing the DNA) of themselves alone. It is *their cells* with their embedded genetic program that combine, divide and multiply to become another per- son. The familiarity here tends to hide the astonishing mystery of it all: a *new individual* is formed from and with the cells and genetic programs of others. The many are *biologically* one.

Offspring differ biologically from their parents in what appear to be only insignificant ways. But these differences build up. Phylogeny, then, the de- velopment of new species, is the accumulation of these "insignificant" mo- difications. In reproducing themselves, things are, however imperceptibly, evolving. Everything living, it turns out, is descended from the same original organisms, built of the same components shaped and formed by the same inner compulsion to survive common to all of reality.

the metaphysical questions

A closer look at this phenomenon reveals that the material substrate itself, i.e., the constituent bits and pieces that are the components of the molecules and atoms that make up our cells, are also completely homogeneous. The foundational building blocks of everything that exists in the world and universe around us are the same for all, as far as we can tell. The elements that constitute the "matter" of our bodies, are exactly the same particles that comprise the protons of the hydrogen fusing in the center of stars. We have to realize there is no reason why it *had* to be like that. But it is. Particle physics has identified them as quarks and electrons. All ordinary matter everywhere, no matter what it is or where it is in the far-flung expanses of the universe, is made up of *only* quarks and electrons — you, me and every one of the 100 billion stars in the Andromeda Galaxy.

All this "sameness" gives rise to metaphysical questions. Where is the **reality** here ... the "being"? Whose cells are these that I call mine, but are made of only molecules, atoms, quarks, ... whose genetic codes ... energy, metabolic processes, mechanisms, traits, abilities? ... whose drive to survive, whose *existence*? Are things many or one? We realize that any attempt to answer those questions puts us on a trajectory into the distant past and to the edges of expanding spacetime that is all-inclusive. Where do we stop? Who or what does *not* belong to the "family"? In tracing my biological genealogy I would be hard pressed to justify stopping at even the first living *prokaryote* cell, which was itself beholden to earlier "pre-living" self-replicating molecules and before them chains of amino acids and proteins. Even the earth itself, from which all these things came, is made from the elements with "heavy" atomic structure like oxygen, carbon and iron — material forged in the heart of giant stars and distributed into space by their cataclysmic disintegration.[50] Approaching the issue from the sequences that produced all matter and by inclusion, all astral structures and all living species, we find ourselves faced with the real possibili-

---

[50] The reference here is to the fact that the "heavy elements" like iron which appear in abundance on our planet are known to be produced *only* in *supernovae*. If such elements exist on earth, it is reasoned, it can only be because the earth (our whole solar system) is the gathered debris of a supernova.

ty that the multiplicity of things might be *only apparent*. **The many may only be one**. What we see before us is a reality shared in both *shape and substance*. This evokes the notion of "participation" but from an entirely different point of view. *The so-called individual entities we see are only the temporary forms of the same evolving substrate and represent the current survival strategy developed in response to environmental pressures.* Everything is constituted of the same "stuff," and elaborates new forms through the same hunger for existence, passing its own special reality on to its offspring eventuating ultimately in new species. *Matter's energy* is at the root of it all: *what* and *that* and *how* things are. The "tree of life," including its pre-life constituents, appears to be something of a mega-individual.

Everything is intimately related. In fact, the genetic connections revealed in the clades of modern science are more comprehensive than what was projected by the older class-based participation. Ancient participation made a collectivity of each separate species but it did not admit any relationship between and among species except for appearances and their common possession of being (and the common origin of their "essences" in the Mind of "God"). In the pre-scientific view, for example, there would have been no possible recognition of a shared reality between the dinosaurs and the birds, because in appearance they were widely disparate and considered separate. But by focusing on the genetic relationships that produce clades instead of classes, modern science has discovered that the birds are the descendents of the dinosaurs. There is more shared reality here than meets the eye and therefore more than the corresponding Greek *ideas* would ever have allowed. So the phylogenetic connectedness unearthed by modern science reveals reality to be even more homogeneous — more "one thing" than the ancients had suspected.

Reality is only one thing? Does this sound familiar? It seems we are back in the conundrum we faced earlier, the *question of the being of the individual and the species which is its reproductive extension.* Given all the sharing — the common substrate and inherited organic structures — what is *real*? ... are individual organisms *real*, am I real as an individual or only a modality of the whole — a tiny leaf on a massive tree? Are we, as individuals, living an illusion? Is there some kind of participatory relationship that supports "reality" at more than one level simultaneously? If the "whole" is not the only "reality" out there, then what grounds the alleged independent reality of individual "things" at another level?   What is the "principle" of their individual reality? Are we all just leaves on a tree?

7. a new world-view, a new metaphysics?

Metaphysics is built on assumptions, axioms and principles that are considered self-evident, but in fact are derived, most often unexamined and unchallenged, from naïve and primitive imagery transmitted by the community. Heidegger's explanation of the Greek categories of *matter and form* based on the image of the "potter" was one example of this process. Our theory in chapter 1 of how the awareness of death functioned to establish the ersatz "reality" of "non-being" in our minds, providing, in turn, a horizon for the characterization of "being," was another. The naïve traditional belief in the ontological individuality of organisms is a third. In our times we have a multitude of new facts that provide new imagery. The discussion just presented of "functional individuality" illustrates the radical impact of those facts and imagery on earlier conceptions of reality and their corresponding metaphysics.

Metaphysics offers an overall conceptual framework that allows us to understand the "reality" that *common sense* has already convinced us is there. In other words, we have to admit we begin with a socially shared common-sense view of things, which we may concede is provisional. In our case, the world-view created by *modern science*, however incomplete its perspective, is our initial horizon and point of departure. It is common sense to us and to our society. We must start there, even if it means that later we adjust our focus. Our philosophy, like everything else about us, is elaborated within the context of our culture.

The West has been wrestling with "reality" questions since the 14th century, when, under the influence of Aristotle, the ancient Platonic world view began to unravel. Science has developed a rigorous methodology and evidentiary requirements to support its new view of reality that has been slowly replacing the ancient view. The culture has been changing steadily through this time. We are the inheritors of that development. It has provided us with an increasing store of solid data in every area of interest. We cannot ignore it.

Some dispute science's accuracy and completeness; they accuse science of skewing reality by the arrogance of its claims to total knowledge. But, there's been a cultural shift. For better or worse, our society no longer doubts the reliability of the scientific approach and method. It has been thoroughly internalized; at this point in time our "spontaneous" perceptions are pre-determined by it. Rejecting it *ad liminem* in the name of philosophy, or morality, or religion (or aesthetics) will, in any event, not eliminate it. The tactic, as we all know from experience, will only create parallel ideologies to which we will have recourse as the need demands. This will nullify any attempt at real intelligibility. The question is rather how do we *integrate* scientific and philosophical knowledge in the pursuit of an integral view of the world. The culture demands that any effective *critique* of science must be launched from within the parameters established by science. I accept that.

philosophical methodology

The general culture has become convinced that the *scientific method,* relying exclusively on observed verification, is the only valid procedure in the search for truth. That means working from the bottom up, not top down, starting with observations and not with principles or assumptions. Out of this point of view came a philosophical methodology promoted by American philosophers William James and C.S. Peirce at the turn of the 20th century that they called *pragmatism,* an unfortunate label in my opinion. Following them, John Dewey modified that and called it, somewhat more aptly, ***instrumentalism.***[51] Its main feature is that it uses something akin to the *scientific method* to ascertain truth. Basically it is a strategy that requires the verification of hypotheses in the practical order (hence the word *pragmatism*). Philosophical analysis is an *instrument* for solving problems, and it is only an effective *practical* solution that verifies that the *knowledge* we have is the "truth."

I plan to use that method as an integral part of the new ***cosmo-ontology*** that I hope will emerge from these explorations, and that will replace the traditional metaphysics. Specifically, it means that we look for formulae, *hypotheses,* that explain the currently known facts. If the conjectures we propose are "true," they should explain all the facts ... and also predict facts as yet unknown.

---

[51] John Dewey *Reconstruction in Philosophy,* NY, Henry Holt, 1920 and Boston, Beacon Press, 1948. Cf also his very lengthy 1916 Introduction to *Essays in Experimental Logic,* NY, Dover reprint, 1954, pp. 1-74, *passim.* Subscription to Peirce's and Dewey's methods and some of their insights does not imply wholesale agreement with their respective visions. In the case of Peirce, that vision was Hegelian *monist idealist,* with which I disagree ... and in the case of Dewey, his **instrumentalism** was intimately connected with a political philosophy that I do not embrace.

This methodology tries to integrate science and philosophy. This is important. The two disciplines must do more than just stay out of each other's way — which is about all they do now. They are not "independent" of one another in any ultimate sense. Philosophers' stereotypical ignorance, if not dismissal, of the verified discoveries of science is unacceptable. Equally stereotypical scientific declarations of absolute reductionism, and that philosophical questions are obsolete, are also invalid.

We accept the current science. It is not only inevitable, it's the way philosophy has traditionally proceeded. The Greeks generated their philosophy to explain their *experience*: a common-sense world-view that held without question; it was science as they saw it. Modern science is the extension and horizon of *our* experience. That means by accepting the results of our science, we are "starting from experience." More will be said about this methodology as its features come into clear focus throughout our inquiry.

This chapter was designed to introduce the various aspects of the "problem of the one and the many" and to establish *matter's energy* as the ruling existential category. We looked at the solution proposed by the ancients: the principles of *essence and existence* and the "participation" relationships created by them and found them incompatible with the discoveries of modern science. Perhaps if we look at the question again from a scientific, cosmological and biological point of view — *from the point of view of matter's energy* — we can deepen our understanding of the characteristics of our reality as it is, and as it functions in the real world of our experience. Perhaps if we can understand what *matter's energy* does, it will bring us a little closer to what *matter's energy is,* and maybe we will come to understand what we are. That is the proposal for the next chapter.

chapter 3
matter's energy
the strategies of survival

preliminaries

We have reassessed the reality around us and seen that it is not shaped and explained by *ideas* existing in another world, the plans in the "Mind of God," but rather by a dynamic material substrate that fills and drives this one. We call it *matter's energy* and it appears to be all there is.

Naturally, then, we want to know what it is about the substrate that gives rise to this complex world we see around us ... for we ourselves are constructed of the same stuff, the same "clay" which is neither created nor destroyed. What exactly is this "*matter's energy*"? ... how does it *do* this? How does that affect me?

methodology

I believe what explains it all is that *matter's energy* is **a living dynamism.** What exactly that means will become clear as we proceed. But in any case such an unorthodox claim will need to be proved. How should we go about that? I propose we lay a solid foundation by *describing exactly what we observe.* What has *material energy* actually done in order to construct this cosmos and ourselves? I believe there are patterns here, strategies, if you will, that open a window on what *matter's energy* is. That very revealing relationship between what things do (their *function*), and what they are (their "nature"), is the utterly simple, uncomplicated premise of a philosophical discipline, using the practical methodology mentioned in the last chapter, that may be called **cosmo-ontology.**

I believe such a discipline is needed to replace *both* traditional essentialist metaphysics *and* mechanistic *scientistic* reductionism.[52] Neither of

---

[52] I use the word "scientism" or "reductionism" to signify a **philosophical position** that holds an interpretation of the relationship of the sciences among themselves and to all knowledge, *viz.,* that

those inherited systems serve us well. The first has been so wedded to an ancient obsolete science as to be virtually unusable in the modern world, and the second, its antithetical reaction, sees the universe as a dead passive mechanism — a view that has contributed to an impoverishment of culture and the fatal degradation of the environment.[53]

These systems of thought appear to be mutually incompatible, but strange as it may seem, I contend that both derive from the *same ancient dualism* of matter and spirit and both implicitly evoke the same dead inertness of "matter" whenever they are employed. The claim that matter is **passive and inert** is required by dualism's prejudice assigning all **life** to an imaginary *essence,* the *idea* of a mind — "spirit." However, if there is no verifiable "spirit," as modern science concludes, it cannot be used to explain life. And *the issue we are confronting in this chapter* **is life in a purely material universe.** In the first two chapters we re-examined and rejected our essentialist-dualist inheritance together with its dominant category of "spirit." In this chapter and the next we will confront reductionist scientism which has stagnated in the belief of the complete passivity of matter. What emerges from our work, hopefully, will be a "third way" that avoids the errors of both these views, and provides us with an understanding of ourselves and our living material universe that is *accurate* and therefore a reliable guide to help us construct our future.

---

reality is fully explained by *being reduced to the science of physics*, and its methodology. It is also called **positivism.**

I am challenging that philosophy, not *science* or scientists. Not all scientists are reductionists. Bernhard Stern, *Historical Sociology*, NY Citadel, 1959, p.15, says, "The doctrine of integrative levels is either tacitly or explicitly accepted by many sociologists and anthropologists ... [the doctrine means] that every level of organization of phenomena has its own regularities and principles not reducible to those appropriate to lower levels of organization, ..." These scientists, however, and many biologists today who claim they are not reductionist, currently do not ground the unique organization of their own science in any principle of intelligibility other than their own assessment of the evidence. In effect they are simply *disregarding* the fact that all reality is composed *only* of the particles and forces studied by physics, which are considered inert. To deny reductionism without seeking an alternative explanation as to what factor might be pushing material behavior beyond the capacities of particles as observed by physics, is to abandon rationality. For either there is a "source" of life that is *other than* the *particles that physics studies*, or the rejection of reductionism must imply the rejection of the conclusions of physics about those particles. Clearly it is not the responsibility of biologists to judge the validity of physics or chemistry. Nor is it of physicists to ground the possibility of biological life and science. But then whose responsibility is it? Without a cosmo-ontology that promotes research designed to ground the vitality of matter, the claim of *non-reductionism* is floating in mid-air, and ultimately non-rational.

[53] Cf Thomas Berry, *Creative Energy*, San Francisco, Sierra Club Pathstone, 1996, *passim.* Berry refers to the absence of the perception of a sacredness embedded the material universe itself.

Reductionism has absolutely no way of explaining life in our universe except to pontificate endlessly that life is simply a reflexive mechanical epiphenomenon randomly produced by the interactions of a dead passive inert matter. What is being proposed in this study is that cosmo-ontology with its inductive methodology can validly show that these claims are simply not true, even without having all the details. Cosmo-ontology doesn't offer a "scientific" alternative, but it sets the issue in a coherent philosophical context — one that encourages science to conduct research in new directions. Matter transcends the limitations once thought to bind it. How it does so will necessarily be a "material" process which doesn't imply inertness. What the actual mechanisms are remain to be discovered.[54]

reductionism and cosmo-ontology

"Reductionism" means every level of organization is explainable by the level below it, ultimately reducing everything to physics. Physics, we must recognize, as an isolated discipline, finds no sign of life and has no reason to look for it. Part of the problem here, clearly, is the artificial separation of the sciences. Since all material reality is constructed of the particles and forces studied by physics, a physics that cannot look at how its "particles and forces" behave later on down the line and up the pyramid of complexity, is cramped by its self-imposed limitations. Effectively it is denied a wider look at its own subject-matter.

Cosmo-ontology picks up where science leaves off. Since the object of our inquiry is *matter's energy* as a substrate, cosmo-ontology looks at reality as a *totality*. Such a perspective *necessarily* **spans the disciplines** and examines connections that the sharp divisions imposed by science's traditional categories cannot. We are interested in *material energy* as *the one* substrate of *all* things *and all* functions *at all* levels.

Cosmo-ontology generates answers in terms that are similar to science insofar as it works from the bottom up, not top down. Events and patterns of events, carefully observed, accurately measured and faithfully reported *by the sciences*, will suggest *interpretations*. These interpretations, in turn, will be formulated as *hypotheses* which will need verification. They will be verified as are all hypothetical propositions, by their ability (1) to explain all the observed phenomena and (2) to predict things not currently observed.

---

[54] cf Stuart Kaufman, *Reinventing the Sacred* 2008, and *At Home in the Universe* 1989, pursues such research.

But at the same time it will resemble traditional philosophy by focusing always on transcendent issues — **all things, at all levels, and always**. It will not limit itself to enunciating sterile statistical correlations without seeking to *understand* the underlying source of their connectedness. The rigid requirements of probity in the physical sciences are important for the accuracy of their work, but limit the scope of the conclusions drawn. Cosmo-ontology, however, will openly employ a methodology that includes **interpretation,** and if there is any bearing, any **sign of living dynamism (intentionality)** at any level, and especially if it should exist at all levels, it wants to know.

[*Author's note:* "*interpretation,*" *and* "*intentionality.*" *These two very significant words are used in this study in a semi-technical sense. Please notice: They are intimately related.* "*Interpretation*" *is a term that refers specifically to the dynamic assessment of a dynamic process which yields accurate understanding. Interpretation is analyzed at more length in chapter 8. It is not "knowledge" as usually expressed by the subject-object relationship, where both the subject and object are considered as fixed and static entities. It is meant as a philosophical parallel to calculus which, similarly, identifies the "drift" of a changing set of relations. It is a tool for understanding process.*

Certain sciences, even now, are essentially dependent upon interpretation for their work. Sociology and psychology, anthropology and ethnology, are not the only examples. What is most relevant for our study here, is that evolutionary biology itself is in large measure the result of interpretations that identify an unmistakable "drift" we call evolution. It's for that reason that evolutionary science has often been accused by its creationist detractors of not adhering to scientific standards of proof. Interpretation, I contend, is a valid, and even indispensable tool of understanding.

"*Intentionality*" *is the object-correlate of interpretation. Intentionality is another metaphorical word for what I mean by saying, "matter is a living dynamism." It is a word that ordinarily and literally connotes consciousness or at least life. Not unlike Bergson's "elan vital" or Schopenhauer's "will," it is meant to capture something of the ancient concept of conatus, or life-force. Spinoza used* **conatus sese conservandi** *to refer to more than the human urge for self-preservation. He says it is the core of all reality. At the human level we can literally call it "intentionality," meaning a "purposeful movement," but at other levels the term cannot be applied literally. It is not a univocal concept. It functions across the spectrum of "things" in a manner proportionate to the complexity present. This analogous application of the term is appropriate because these dynamic processes* **exhibit similar behavior and have the same existential import** *despite occupying different levels of emergence. This is a matter of valid scientific interpretation. I believe this will become clear as we examine the concrete evidence.*]

There are two questions that cosmo-ontology deals with that science ignores, or even actively rejects: (1) *what things are,* and (2) what "*inten-*

tionality" they bear, i.e., *what they "want."* In chapter 2 we saw that the first question, "what things are," had been traditionally answered by the ancient dualist theory of essences or natures *as a function of teleology, purpose* — each thing was believed created by "God" and given an "essence" *with a purpose.* We rejected that theory; evolution proved it was no longer tenable.

The second question, about *intentionality, means that we are **looking for the precursors of life.*** Such a quest has always been dismissed by *scientism* as a foot-in-the-door for "creationism." But it is not necessarily so. If we come at the issue in reverse order as we have suggested, and let function — the behavior we observe — tell us what possible "nature" explains it, then the nature it reveals will be one which determines function, not one that is used to explain a gratuitously imagined purpose.

1. life and integration, the "nature" of *existence*

L et's begin building our hypothesis.
I believe the observations made by evolutionary biology recapitulate what is seen by the particle physicists despite reductionist denials. The general phenomenon can be described this way: "the attributes of life were at some time evoked in inanimate matter by the ***action of a force*** of whose nature we can form no conception."[55] ... and: "we may fairly infer ... that there is probably in nature ***some agency*** whereby the complexity and diversity of things can be increased."[56]

These general statements are not offered as "science" but as an initial articulation of an hypothesis to be explored. While they remain indefinite, what is important for our purposes is that they are *neither* attributing vitality to an *immaterial reality* like "spirit" or "essence" existing alongside matter; nor are they saying that life can be explained by the random mechanical interactions of an inert matter. They speak of a "force" or "agency," which suggests that life is radically a *property of matter itself.*

existence is a one-way street

We may tentatively say, prompted by observers like those just quoted, that the universe is the progressive unfolding of one homogeneous sub-

---

[55] Sigmund Freud, *Beyond the Pleasure Principle*, NY Bantam, 1919 / 1959 p.71
[56] C.S.Peirce, "The Doctrine of Necessity Examined" in *The American Pragmatists*, Konvitz and Kennedy, eds. Cleveland, Meridian Books, 1892 / 1960, pp. 138-9

strate (*material energy*), full of life, development and increasing complexity.

Random mechanical interactions will as likely move in one direction as in another. An accumulation of elemental particles at the most primitive level, for example, resulting from purely random interactions *without any other factor in operation*, will as likely reverse direction and dissolve. One would expect the universe to be a vast structureless soup of swirling, unattached energy. But it is not. It is a complex, structured, highly developed and inter-related elaboration of one single "thing," *material energy*, the exclusive substrate present in every form that reality takes. Its increasing complexity — what we may call, metaphorically, "growth" — never reverses; given the right conditions, it moves in only one direction: ever higher complexity through time.

I am going to suggest a preliminary hypothesis based on this sketch. It has two parts. The first part is simple enough: **matter is a living dynamism.** This is hardly new. It's virtually a corollary of the premise that rules our explorations — that there is no "spirit" that is a separate genus of being — *existence* is *material energy*. Since life abounds, it follows that *life's dynamism must be an intrinsic property of matter's energy.*

I offer two brief supporting arguments for this first part. First, if it were not, life would have to have been inserted by *something other than* matter. That could only be something immaterial, which is, by our premises, impossible because what is immaterial does not exist. Second, if life were not an intrinsic property of matter, once inserted from the outside, it would have to be *continuously sustained* by an outside immaterial source, the same or another, for matter would not possess the wherewithal to sustain life on its own. All life would have to be *continually maintained* by something outside matter. But there is nothing outside matter, therefore ... etc.

*[**Author's note:**   All these words, "life," "growth," like "intentionality," when applied at levels below the biological are **metaphors**. They are words taken from existence as we experience it at the level of organisms and consciousness. They do not apply in the same way in every case, and certainly not to pre-living entities. Calling the phenomenon **conatus sese conservandi** as Spinoza does, gives it pretensions of being technical, but it is, in fact, no less metaphorical. **Conatus** was originally coined in ancient times to describe living things.*

*But, despite being metaphors they all refer to something **real** whose description at this point cannot be given in mathematical or scientific terms. These metaphors are all we have. But they must be taken in proportionate measure. They mean something real but they don't mean the same thing at every level. They are uniquely* *expressive of the sense of drift, direction, bearing, intentionality — a creative existential dynamism whereby matter's energy moves against entropic dissolution to develop ever more survival-efficient forms. There is a "lowest common denominator" that characterizes what existence does at all levels — it is a self-embrace. Life is built on pre-life events and operators that moved toward coherence. It is a non-conscious or pre-conscious* **existential self-embrace.** *That phrase comes closer to the reality, but it is still a metaphor. The historical survey we are about to conduct, hopefully, will clarify the issue.*]

The second part of the hypothesis is not as simple: The world is full of complexity, and not only among living things. Our hypothesis is that *the pattern for maintaining self-identity is universally and cumulatively communitarian.* This "strategy" is utilized at *every level* of emergent existence; it is not only a biological phenomenon. *The existential self-embrace, survival, depends upon a* **progressive aggregation of constituent elements — integration building on earlier integrations.** The complexity which accompanies the unidirectional development of living organisms is the product of *a process* that, over and over again starting at the time of the big bang itself, *cumulatively* **constructs one thing out of many.**[57] Growth, an expanding range of function and increasing organic complexity go together with the production of *species,* i.e., a potentially limitless community of genetically related individuals.[58]

Let's flesh out this hypothesis with a survey of the astonishing accomplishments of matter's energy.

[**Author's note:** *Please be advised, neither the existential (survival) character of evolution nor any of the remarkable facts we are about to survey is intended as a "proof" for the existence of a traditional provident designer "Spirit-God-person" or even of the "anthropic principle" that some use for that purpose. "Spirit" is a theory of essentialist metaphysics. Insofar as we have rejected essentialist metaphysics as untenable, there is no possibility of anything existing "outside" our world of material energy and intervening in it. Whatever happens is a function of material energy. Whatever "God" there is, is, or is made of, material energy.*]

---

[57] and also creates many things out of one ... through reproduction

[58] This is not "supernatural," it is a feature of all systems, including mechanical. Scientists working on robotic systems concluded: "*When compared to the evolution of networks with fixed structure, complexifying networks discover significantly more sophisticated strategies,* " Kenneth Stanley and Risto Miikkulainen *Competitive Coevolution through Evolutionary Complexification,* Dec 2002, Technical Report UT-AI-TR-02-298 DCS, The U of Texas at Austin. In another study, multicellularity "*... also shows emergent self-repair, which is used to produce highly resilient organisms.*" D.Federici & Keith Dowling, "Evolution and Development of a Multicellular Organism" *Artificial Life* (July 2006) pp 381 - 409

big bang nucleosythesis

We start at the beginning of time.

Science is able to reconstruct events in the formation of our universe down to portions of a second beginning with an energy eruption of unimaginable magnitude that occurred 13.7 billion years ago called universally by scientists, the "big bang."[59] That event involved *all the material energy* that exists in our universe, and began an accelerating expansion that has resulted in the formation of 100 billion galaxies, *each one* containing an incredible 100 billion stars. Our own galaxy is 100,000 light years in diameter, and is 2.2 *million* light years from its nearest neighbor, the galaxy Andromeda.[60] If we take these distances and numbers as average, it gives us some idea of the size of the universe. There's no sense looking for a descriptive adjective ... it is an immensity that is simply beyond words.

While there is a great deal known about the ordinary matter (protons and neutrons) formed in the big bang, the nature of the energy that preceded it (and may still exist in the form of dark matter and dark energy)[61] is a matter of conjecture. Some have hypothesized homogeneous loops ("strings") of energy whose variations in vibration account for the differences in the particles and forces — the quarks, electrons and gauge bosons (force-fields that sometimes act like particles) — that are never found alone but are the constituents of the fundamental particles that make up ordinary atoms.[62]

What is clear is that *from its very inception* **ordinary matter is structured** ... by that I mean that it is complex, it is a collectivity; it is formed from aggregations and integrations of "smaller" components or constituent parts. The quarks and gluons ("strings"?) are the simplest elements identified so far. These quarks / gluons are inherently *unstable* i.e., they were an amorphous energy plasma until they themselves *aggregated and inte-*

---

[59] For all the facts relevant to this section, see Michael Turner, "The Universe," *Scientific American*, Sept. 2009, pp. 36-43

[60] Light travels 6 trillion miles in one year.

[61] CF Sean Carroll, **"Dark Matter and Dark Energy,"** a course produced by *The Teaching Company* 2007, lecture 1, "The dark Side of the Universe." These are inferred to exist because the gravity calculations of the observed rotational features of galaxies and clusters of galaxies demand the presence of matter that cannot be seen, and the accelerating rate of universal expansion cannot be explained without positing a suffusive energy that has yet to be identified or detected.

[62] Brian Greene, *op cit. passim*

grated as the protons and neutrons that are the constituents of atoms, an event that occurred ***within the first second of the big bang***.[63] These particles congeal and take on their physical character only in the act of baryon formation.[64]

Atoms, which are built of protons and neutrons, are stable. They are *communities* of particles that provide the most primitive stability in our universe, as far as we know. Most of the ordinary matter of the universe is in the form of Hydrogen atoms, the simplest atom, one proton.

[***Author's note:*** *One of the discoveries of particle physics that is most difficult to imagine is the existence and function of **gauge bosons**, which are force fields that sometimes act like particles. They recapitulate the well-known inconsistency about light (and electricity) which, similarly, sometimes acts like a wave (magnetic field) and sometimes acts like a particle (photon or electron). The **gluon** is one of these gauge bosons, and a very important one because it is responsible for what is known as the "strong nuclear force" which binds the quarks together to form protons and neutrons, and then binds those baryons together to form atomic nuclei. There would be no "ordinary matter" and therefore no structure in our universe without the existence of a "force" to bind the interior of atoms. The binding that occurs does not function at extremely high pressures and temperatures. In conditions like those prior to one second after the big bang (achieved in the latest particle colliders) quarks and gluons become free and an energy plasma forms.*[65]*

Are quarks and gluons distinct independent entities, each with their own properties — gluons with the property of a force-field, and quarks with the property of a particle? String theory says "no." It says they are all strands of amorphous energy which, when they "vibrate" in different ways, assume different properties. But the question remains, why do they vibrate as they do? Are they **determined** beforehand to vibrate this way or that ?... (some suggest there are up to 11 "dimensions" in space, which force energy strings to spin and vibrate as they do)*[66]* ... or do they, in the presence of one another **spontaneously** assume the spin and vibration necessary for one to form a quark and another to perform the function of a gluon so that protons and later atoms can be formed? And if so, what explains this interactive spontaneity and why does it "prefer" the formation of baryons (protons and neutrons) to remaining a free plasma?*]

Thus even at the most primitive levels currently known, the patterns of *aggregation and integration* seem to be the condition for achieving stability and identity. We will pick this up again shortly. What it means is that

---

[63] Carroll, *op.cit.* "Dark Matter, Dark Energy" lecture 11 "Big Bang Nucleosynthesis."
[64] **baryon** is a term for sub-atomic particles constructed of three quarks, like protons and neutrons.
[65] cf report on quark-gluon plasma (QGP) on the RHIC website maintained by the US Dept of Energy, Office of science: bnl.gov/RHIC/default.htm. The Relativistic Heavy Ion Collider (RHIC) is maintained at Brookhaven National Laboratories.
[66] Brian Greene, *The Elegant Universe*, NY, Vintage, 1999

from our point of view, anything at all that we can experience — **anything** — exists by reason of this intrinsic, inescapable fact: *it would not be here if it were not complex and structured, i.e., cumulatively composed of multiple elements, the result of a communitarian process of aggregation and integration. Matter is a collective phenomenon.*

As the universe expanded and cooled, stable hydrogen atoms (with one proton) fused to produce helium (with two), releasing tremendous energy. Fusion is what occurs in a hydrogen bomb. It is also the process responsible for the formation and activity of stars, like our sun, whose enormous energy output, effectively *billions* of hydrogen bombs converting continuously into helium through self-fusion, has lasted for billions of years providing our earth with an enormous amount of free energy. Some of these stars are massive enough to permit the coming together of even heavier atoms (containing more protons and neutrons) like carbon, nitrogen, oxygen, phosphorous and other elements ... terminating in iron. Together they fill out a periodic table of the elements which display such linear regularity that the gaps that appeared at first, accurately predicted the existence and character of elements that were later found.

These massive stars, however, are short-lived.[67] They "quickly" (by astral standards) explode in what are called *supernovae* and spew their contents into the surrounding space in immense clouds laden with these heavier elements. It was out of such enriched debris that our solar system, a second or third generation sun and its planets, formed 4.5 billion years ago. It explains why our earth has so many of those elements which are essential to life.

increasing complexity through integration

Fast forward to life on our planet ... the ability to aggregate, integrate and complexify is most clearly visible in the development of the community of living things on earth. This is not a recent phenomenon. Looking back on the evolutionary history of life on earth, it began with the formation of **complex molecules** which are the aggregation and integration of various atoms (carbon, water, oxygen, nitrogen) to form entirely new substances with entirely new properties.

---

[67] Most stars live between a billion and 10 billion years. Massive stars (more than 8 solar masses), however, live only about a million years.

Experiments with basic non-living elements (methane, ammonia, water, hydrogen) gathered to simulate the earliest (non-oxygen) atmosphere of the newborn planet earth, when shocked through with electrical current, showed that "biologically important small molecules, including amino acids, sugars, and significantly, the building blocks of DNA and RNA, ***spontaneously form*** ..."[68]   Other experiments with *autocatalytic molecules* have shown their ability to replicate themselves and even faithfully to replicate their variations.[69]

This process occurring over a billion years hit upon two molecular combinations that were key for the development of life. The one produced **nucleotides** that connect to form RNA which is able to replicate itself and pass on genetic instructions ... the other, **amino acids** that are the segments that link together to make proteins which are the structural elements in all living cells.

Viruses use RNA molecules and may have been a key steppingstone taken by *matter's energy* on its way to life. "Viruses cannot be called living."

> They [viruses] have been described as organisms at the edge of life, ... they do not have a cellular structure, ... [they] do not have their own metabolism, ... they cannot reproduce outside a host cell ... . Accepted forms of life use cell division to reproduce, whereas viruses spontaneously assemble within cells, which is analogous to the autonomous growth of crystals. Virus self-assembly ... lends further credence to the hypothesis that life could have started as self-assembling organic molecules.[70, 71]

Life emerges

There is fossil evidence that life began on earth just as soon as things had cooled down enough to support it in its most primitive forms. Even though modern science believes it knows the various "steps" leading up to life, the exact "agency" or "action of a force" that was the vehicle for the

---

[68] Richard Dawkins, *The Ancestor's Tale*, Boston, Houghton Mifflin, 2004, p.566

[69] Ibid., p.572. Note also, theories that life may have entered earth from debris ejected by passing comets or volcanic activity on Mars support the contention that the spontaneous development of life will occur *anywhere conditions permit it.*

[70] Wikipedia, *Virus*, http://en.wikipedia.org/wiki/Virus as of June 29, 2009. Cf David Goodsell, *The Machinery of Life*, NY Springer Verlag, 1993, p.112.

[71] Another theory for the "place" of viruses among the structures of *matter's energy* was offered in private correspondence by biologist Francis X. Lawlor. He suggests that the viral strategem of tricking a normal cell into reproducing the virus — a roguish ploy indeed — was the development of bacteria ... it postdated the emergence of life and represents a regression into a semi-inanimate state that was "selected" because of its incredible survival efficiency.

transition from non-life to life is not known. This becomes a problem for the reductionists because it appears that all the forces and components are known. If there is another "agency" responsible for life, it would have to be in an area that is not yet accessible to science — between or behind or underneath what is known. But no one really suspects there is anything "new" to find. Current efforts are directed at exploring "complexity and chaos" theory. The aim is to ascertain whether the known forces and elements might possibly reach a "tipping point" of volume or intensity of mix beyond which the previously "inert" elements spontaneously "come alive."[72]

The point to emphasize here is that in such a case the "agency" discovered is not really something new but rather a set of circumstances that permits *something already present* to display a property possessed all along. Life, in other words, is not a "property" of the "agency" (complexity or chaos), it is, rather, a property of *matter's energy* that under the proper conditions *emerges* and becomes functional. Something present but dormant begins to function and become perceptible — an intrinsic potential self-activates. This is precisely what our hypothesis is claiming. Life arose 3.8 billion years ago when those conditions were met.

Cyanobacteria and other single-celled organisms known as *archaea* have left fossil records that date as far back as 3.8 billion years. These most primitive of all protozoa, like present day bacteria, were characterized by the absence of a nucleus and reproduced by simple fission. Yet they were themselves communitarian integrations made of and utilizing *thousands of complex molecules* for their structure, metabolic processes and reproduction. They are collectively known as **prokaryotes** and for a very, very long time (almost two billion years) were the only form of life that existed on earth, as far as we know. True to the patterns emerging in this survey, however, these primitive unicellular organisms found ways to *join together in communities* with one another that increased the possibility of survival, the range of activity and reproductive success.

> Intercellular communication and multicellular coordination are now known to be widespread among **prokaryotes** and to affect multiple phenotypes. ... Bacteria benefit from multicellular cooperation by using cellular division of labor, accessing resources that cannot effectively be utilized by single cells, collectively defending against antagonists, and optimizing population survival by differentiating into distinct cell types.[73]

---

[72] Kaufman, *op.cit.* passim.

Eukaryotes. And then there was the stunning innovation 2 billion years ago of the **eukaryote** cell. This larger and much more complex single celled organism, the inventor of sexual reproduction, *emerged* out of the primitive collectivity of prokaryotes almost 2 billion years after life appeared on earth. Unlike the prokaryote, the eukaryote cell displayed a developed nucleus and a protoplasm containing membraned organelles — energy "power plants" known as *mitochondria* in animals and *chloroplasts* in plants — that were themselves the *result of communitarian aggregation and integration*. For, what is remarkable about these organelles (often called cellules), is that they are *unmistakably prokaryotic* in structure and function. The universal interpretation of this phenomenon is that the eukaryote was an innovation that emerged as a product of *endosymbiosis*: a process by which one cell is taken up by another and retained internally, such that the two cells live together and integrate.[74]

The eukaryote cell, apparently, is *emergent*; that means its structures and activities are not explained by the far more limited prokaryote cells from which it evolved. In that evolution, prokaryote cells lost their independence as individual living organisms in order to be incorporated into a more efficient, more wide ranging and more survivable *new living community* managed by the eukaryote host. The erstwhile prokaryotes were totally absorbed into the new cell, not as the independent participants of a colony, but now rather as integral components of an entirely new phenotype. But we know they were once prokaryote cells because even now they have no nucleus and besides their own membranes (containing their own separate DNA) dividing them from the host's protoplasm, they retain their own nutritional and reproductive patterns which are *unmistakably prokaryotic* and *not* eukaryotic. They produce the energy macro-molecule known as ATP, for the benefit of the eukaryote host, using their inter-membrane space, which is exactly what all prokaryote cells do, including the bacteria of today. And when the eukaryote host cell divides in the reproductive processes called mitosis (asexual) and meiosis (sexual), *the mitochondria proceed separately to replicate according to the more primitive prokaryotic pattern of binary fission.*[75]

---

[73] James A. Shapiro, "Thinking About Bacterial Populations As Multicellular Organisms," *Annual Review of Microbiology*, Vol. 52: 81-104 (October 1998). The quote is from the Abstract.
[74] Margulis, L., *Symbiosis in cell evolution*. W. H. Freeman and Co., San Francisco, (1981)
[75] Goodsell, *op.cit.*, pp.65-80.

Sex. Sexual reproduction was an invention of the eukaryote cell. Ursula Goodenough elaborates:

> Eukaryotic sex was both ancient and ubiquitous. It arose some time prior to the Cambrian explosion, and is found in all the phyla that trace back to the Cambrian. ... The origin of sex marks the onset of **biological relationship** — as contrasted with the solitary asexual existence of the bacteria and amoebae ... Once procreation was handed on to germ cells, and embryos and offspring, their protection assumed vital importance and, in animals, was entrusted to strong emotional instincts ...[76] (emphasis mine)

The implications seem clear. Sex was itself an application of the communitarian strategy in a most sophisticated way that enhances survivability. What we know as "family," the basis of our psychological and sociological characteristics as a species, that which is unique in determining not only our personal life, but our very existence itself, finds its etiology in the eukaryote cell and from there is passed on to every life form on the planet. The significance of this can hardly be overstated. Sex, which absolutely dominates our human existence and personal destiny — the source of human energy, passion, creativity, as well as pathology — was the creation of a microscopic one-celled animal born in the sludge of a tidal pool *two thousand million years ago.*

the cambrian explosion

Multicellular organisms have arisen independently a number of times starting from about 1.5 billion years ago when the atmosphere contained sufficient oxygen to support the higher energy requirements for multicellular life. But most salient for our survey was the discovery on the part of eukaryote cells that they themselves, in turn, could cooperate in communities and build **multicellular individuals,** — animals and plants as we know them. This re-application of the strategy of aggregation and integration, in the hands of the high powered eukaryote cell with its more sophisticated capability, gave rise to the Cambrian explosion 550 million years ago and resulted in the rapid development of all the phyla of living things that populate the earth today. The eukaryote's greater energy and expanded range of activity was quickly exploited for the creation of specialized organs and structures. *All complex multicellular plants, fungi, insects and animals, unmistakably new independent species and not colonies, in-*

---

[76] Goodenough, *op.cit.* p. 117

*cluding humans, are developments constructed exclusively of **eukaryote** cells.*

The subsequent evolutionary history of complex plant and animal life on earth is familiar to us all. Organic system after system was discovered, submitted to the judgment of "natural selection" allowing for ever-greater scope of living activity and adaptation to more environments. The stunning panorama of living things that inhabit every possible environmental niche on earth is the result of *the repeated use of the communitarian strategy of aggregation and integration* of **eukaryote** cells for the enhancement of survival, the expansion of range and versatility, leading to ever greater diversity and a wider community of offspring.

All of this development is time-related; the most complex come later precisely because they are *cumulative* effects built on and from earlier constructions. Every evolutionary success was an integration dependent upon earlier integrations. This is too prominent a feature of our biota to be ignored, or even minimized. We are looking straight in the face of a living spontaneity. *Matter's energy* appears to be one growing thing ... almost like a tree. The branching cladograms we saw in chapter 2 reflect reality.

Our survey shows that communitarian aggregation and integration is a consistent and repeated pattern in the emergence of everything we can see in the universe ... in the heavens and on earth:

- the assembly of protons and neutrons from the integration of quarks and gluons which emerged out of primeval homogeneous energy;
- the congealing of hydrogen atoms from protons and electrons, accounting for the mass and energy output of the galaxies and stars;
- the construction of complex heavy atoms combining simple ones, eventually creating the elegant periodic table of the elements;
- the formation of molecules from combinations of atoms;
- the coming together of highly complex as well as self-replicating molecules from simple ones;
- the emergence of unicellular prokaryote life constructed of and fed by a manifold of complex and self-replicating molecules;
- the formation of interactive cooperating communities of primitive prokaryote cells;
- the emergence of the eukaryote cell as itself a *community* formed from the aggregation and integration of prokaryote cells;

- the invention of sexual reproduction by eukaryote cells expanding genetic relationship through cooperative collaboration;
- the formation of multicellular and multi-faceted organisms from the integration and cooperative specialization of eukaryote cells. ... all progeny of the same parent.

Appendix 8 contains a chart showing how the growth of molecular complexification went together with the major stages of macro-evolution. *These novel recombinations were effectuated by the living organisms themselves* as they forged the very molecular infrastructure they needed to survive. What is noteworthy is the continuity of this process across the life-divide. Molecular complexification followed the same patterns and built on the same infrastructure in life as in pre-life, suggesting the same forces were at work.[77]

This is clearly a new way of looking at the *one and the many.* But how does understanding this astonishing phenomenon impact humankind?

cells and the human self

There are trillions of living cells that make up each *human organism.*[78] These cells, we are now learning, are also living individuals with their own independent history. We tend to think of them as *human cells* as if they were created by our bodies, and in a sense they are. But in another sense as we've seen, they are not. At a level far more fundamental than their human functions, all human cells are **eukaryotes.** We understand now what that means. Our cells have a biological structure and mode of operation that is historically traceable not to their role in the human organism, but rather to the creative advances made 2 billion years ago by a tiny animal, the eukaryote unicellular organism. *All complex plants and animals existing today and for the last 550 million years are integrated aggregates of eukaryote cells.* You and I are their direct genetic descendents.

The relevant point here is that these eukaryote cells, *shared by every multi-celled life form on the planet past and present,* have a structure and function that antedated and is *independent of* the organisms that make use of them. Our human organisms had no hand in their discovery, basic design or operation. This bears emphasizing: *the cells that comprise my*

---

[77] Libb Thims, *The Human Molecule,* LuLu.com, 2008 Cf *Institute for Human Thermodynamics,* URL http://www.humanthermodynamics.com/Evolution-Table.html
[78] Estimates range from 50 to 100 trillion.

body, in a very real sense, are not mine. They are not even exclusive to human organisms. They *have a life and history of their own.* I am dependent upon the ancient metabolic and reproductive innovations of these living entities which are responsible for my own "personal" needs and abilities — nutritional, metabolic, reproductive, emotional, intellectual. I am biologically limited to and driven by their organic potential. I can do only what their capacities allow me to do.

Nor does their cooperative role in my organism guarantee the complete elimination of their independence. My cells' capacity for autonomous activity is on lethal display in the cancers which seem to be attempting to cast off the yoke of collective submission and return to primitive self-determination.

But cancer is an aberration. The living organism is constructed on the principle of absolute cooperation. Complex organisms would be impossible if DNA could not rely upon a lock-step obedience from the cells used for their construction and function.

We conclude that the two living "individualities" that co-exist — the cooperating cells and the resultant integrated organism, in this case, ME — are not *two "things."* They related to one another *as phases of the same biological event.* They are the sequential "apparitions" of *one* ongoing *process*, one individual, the unfolded outgrowth of eukaryote cells. This is the *integrated function.*[79] It is simultaneously them and me. I am the eukaryote cell in its latest and greatest strategy of survival.

summarizing the survey

Before moving on, let's see what our survey suggests. The big bang itself has been described as an improbable event establishing a *material universe* of such low entropy (high potential energy) that all events thereafter could occur "downhill" utilizing energy and accumulating entropy. The big bang is indistinguishable from the near instantaneous (within one second) formation of protons and neutrons — each of these baryons being a collectivity of three quarks and three gluons.[80] It could be described as *the* primordial communitarian event, setting a foundational pattern that has been repeated endlessly in a cumu-

---

[79] *Integrated function* is a term for what Aristotle might have called "substances." Ken Wilber would call them *holons* but he uses that term more broadly to refer to any and all manifestations of integration — not only biological but logical, psychological, etc.
[80] protons and neutrons are baryons

lative manner to form a vast pyramid of related evolved structures, the earlier ones seemingly passive and inert, the later ones undeniably *living participants* in *existential construction* (evolutionary survival). All build on and repeat the communitarian patterns of aggregation and integration established in the first instants of the big bang, and all in function of continued existence.

It seems sufficient and necessary to distinguish between some absolute victory over entropy and the identification of an **anti-entropic dynamism** that — however pathetically temporary — is a clear manifestation of the presence of a dynamism that **grasps at existence** at all levels ... wherever *matter's energy* is found. I am not concerned to prove that the **conatus** actually accomplishes its goal ... my only interest is (1) that the dynamism *does exist*, (2) that it exists at *all levels* even the most primitive, (3) and that it reveals *material energy* to be precisely an *energy for existence* ... an active, even **defining** self-embrace, regardless of its apparently tragic failure to achieve endless existence. *Existential energy — which produces the conatus —* doesn't just accompany *matter,* it's not just a property of matter, *it is matter.* The very "nature" of *matter's energy,* in other words, *is to exist.*

Non-living, self-replicating molecules have been concocted in the laboratory. It is believed they occurred naturally in the distant past and were a key steppingstone on the way to life.[81]

Granting that self-replication was a random discovery of molecules, why wasn't the entropic dissipation of structure and energy selected? And when you take a step back, why has this happened repeatedly — *every time, in fact, there was the opportunity?*

Replication, especially in the form of genetic reproduction, trumps death and dissolution by creating new individuals. This results in a stable continuity despite the fact that individuals disappear. I don't want to give the impression that "selection" is a *purposeful* choice ... at this level it is not. "Life" is not necessarily "conscious" even though in the most primitive biological forms like prokaryotes, it shows the ability to respond autonomously to "information," — survival-related stimuli — moving toward food and away from enemies. And later, even at the human level, life is not

---

[81] Mallove, Eugene, "Self-Reproducing Molecules Reported by MIT Researchers." *Tech Talk,* MIT News Service, May 9, 1990. Cf also the recent article by Ricardo & Szostak "Life on Earth," *Scientific American,* Sept. 2009, pp. 54-61

always "rational," but it is certainly not ever passive and inert. It actively promotes itself and fights to maintain the integrated structures it has fabricated.

What we see in all the structures of the universe starting with the most primitive and inorganic, is that **existence is a self-embrace.** *Matter-energy* spontaneously counteracts the random forces of dissolution and dispersal by cumulatively building on prior counteractions, *wherever conditions permit it to do so,* in order to sustain collectively achieved coherence.

The existence of self-replication in *non-living molecules* is important because it forms a bridge between life and non life. Self-replicating molecules are even more primitive than viruses because they do not have RNA (as viruses do) to encode a genetic program, and so they are not said to be *alive* in any sense of the word. But for that very reason they confirm *what matter's energy does whenever possible, wherever it is found,* to the maximum of its current potential. This is what we mean when we say "matter is a living dynamism." It is an existential self-embrace.

the *analogy* of processes

What I'm calling "life," is *not a univocal notion*; it does not manifest exactly the same behavior at all levels. Nor is it exclusively a feature of "things," identifiable entities, organisms. It is a *generic label* that refers to a property of the "nature" of reality. *Material energy* is a dynamism which is creative of life. "Living dynamism" is meant to describe the *process* functioning in both the integrating unit *and the surrounding environment,* proximate and remote ... *all* of which must collaborate to allow for stable *identity* to exist and perdure. It is a common feature of our universe ... absolutely determinative of complex development. It is the very definition of *existence.* Matter "wants" to *be-here* and enters into bound relationships, i.e., *collectivizes* to do so. **Matter's energy is an existential dynamism whose communitarian self-embrace creates, constitutes and, therefore, explains all things.**

Selection, which is essential to this process, is a feature of the whole, which includes the entire environment proximate and remote. It explains why it would take hundreds of millions of years for the first living organism to emerge ... and thereafter, thousands of millions of years more for complex multi-celled life-forms to arise. The emergence of life was a project and selection of the Whole, not the adventure of a part or the whimsical insertion of an outside "Creator."

2. individuality — an interior view

Get ready to change gears.

Up to this point we have been looking at the issue of integration — which may be called *the* communitarian strategy — from an evolutionary biological point of view in the pursuit of verifying an hypothesis that states that *matter is a living dynamism that spontaneously collectivizes and complexifies.* I would like to take an entirely different approach in this section and try coming at it phenomenologically, — from human self-consciousness.

Because of the nature of our self-consciousness, we can look at our organic communitarian individuality by direct introspection. I would like to take advantage of this privileged window of observation. Now that we understand a little more about the organic infrastructure of our personal lives, what more can direct introspection tell us about this phenomenon of biological integration? Remember: we are still fleshing out our hypothesis stated at the beginning of section 1 of this chapter.

When we first considered the question of individuality in chapter 2, it was suggested that modern *scientism* would be quite comfortable with the view that the individual is "only" a *modality* of the whole and not an onto-logically independent "something." Given the purely functional interests which characterize science, it does not seem important that anything below the level of the whole universe of *matter's energy* be identified as an independent existing *substance*, a "thing."[82] But we shouldn't overstate this. As we saw in chapter 2, scientism doesn't deny the "metaphysical" individuality of the organism; it just doesn't care one way or the other. It is interested in other things. A functional individuality is as far as science needs to go.

Our thumbnail sketch of the history of universal development was based on the evidence uncovered by science. The diversity of form and function is constructed from the *same substrate,* using the same repeated patterns of integration initiated at the big bang itself that result in a single

---

[82]A "substance" for Aristotle was an individually existing instance of an "essence." It was claimed to be the locus of *ontological autonomy.* Spinoza said that all of Nature was one single substance, "God," because only "God" had such autonomy. All things, he said, were mere modalities of "God" because they existed "in" God. Bernard Lonergan calls substance a "thing." We should not be surprised that modern science, which does not recognize "essences," might not recognize "substances" or "things" as autonomous individuals either.

genetic "family." Might the material universe be a single organism? What would that mean for me as an individual human being?

The Buddhist doctrine of *anatman* (no-self), might be understood to say that the human individual is only a modality of the whole. The Buddha's admonition to deny the self's clamor for aggrandizement is justified by "metaphysical" suggestions to the effect that separate entities do not exist — which includes the human "self."

We in the West, however, traditionally are not comfortable with that perspective. We are invested in the importance of the *individual* because we each consider ourselves to be *a person* with a separate destiny as preached by our religion. But we must recognize that we do so based on an ancient metaphysical hypothesis: the independently subsisting immortal soul. Given that belief, it is no surprise that the sense of self in our culture has been intensified to the point where it has become the centerpiece of the Western view of the world. If we in the West are obsessed with individuality and saddled with the political and economic structures that derive from that obsession, I believe, this is where it comes from.

Science, however, challenges this view, at least in theory. It says that it cannot verify the existence of a separate human "soul." If the lack of verification really means that there is no uniquely created immortal soul, does it also call into doubt the metaphysical independence of an individual "personality" which was based on that traditional understanding? Is the self, the person, an illusion?

the "separate self"

The traditional belief receives what seems to be empirical confirmation because internally we experience ourselves as independent realities unaware of any connection with the material foundations of our personalities. We are totally oblivious to the many trillions of living cells divided into specialized functions which comprise our organs and manage our well-being. This is an important and instructive phenomenon which we will take up again shortly. Here I just want to emphasize that this *natural structural ignorance* explains how it was possible for people to have believed either in the independent existence of a spirit-person divorced from the body, as in the West, or as in India, that the "self" is an illusion altogether.

But the phenomenon of "self awareness" is biologically based and explainable without recourse to either of those fantasies. Ironically, we know it is the very cells of the body — the neural and hormonal networks sustaining consciousness — that are responsible for the disembodied experience of personality. How do we know that? Very simply because if parts of these systems are damaged or destroyed, the experience of self and other "spiritual" functions are seriously modified, if not eliminated. Even the very sense of who I am, is dependent on the integrity and health of the organic systems that support it.[83] Self-consciousness is not the product of a separate "spirit," nor of our imagination; it is the product of the material substrate, the specialized cells of our body, our "flesh."

Personality, identity, the sense of self, is the expression of an *integrated function.* My body is not an *ad hoc* temporary federation of independent cells. The cells are one organism. They provide the specialized structures and perform the functions of the one entity, internally experienced as *one thing.* In my case I call it *me.* This is identity.

What is "identity" after all? At the risk of the distraction of another change of gears, we need to add a long parenthesis here and try to analyze this notion. It's central to everything that exists. Just how should we understand *identity*?

Identity, process and time

We can start by saying: *identity is a function of time.*

Going back to the "big bang," at *material energy's* most primitive level, baryons represent an initial stability within the "quark-lepton plasma," the roiling energy of the initial expansion. What does "stability" mean? It means **identity** — the related particles in question **remain related through time.** They do not deteriorate into their constituent elements, they do not morph into other forms or back into the raw energy of quarks or the vibrating strings thought to constitute them. *Identity (stability) is the perdurance-in-time of matter's energy in a form acquired by its communitarian intra-relationships.* Unattached and unrelated energy achieves *identity,* as far as we can observe, *only* by entering into bound relationships

---

[83] The claim that there is a "soul," a principle of personality that is "*different from*" the material substrate even while being *totally dependent* on it, can only mean the existence of an immaterial genus of being (reinstating a dualist universe) ... which in this case would be functionally irrelevant because it is acknowledged to be *totally dependent on* the organism for its exercise. It could not function after death because there would be no organism for it to depend on. So to insist on its existence as a "separate" genus of being *only while alive and totally dependent on the body* is purely academic. Even if it were true, it has no practical import whatsoever.

that endure through time. It's the communitarian process that we have been observing. It is, apparently, not avoidable.

Time is intrinsic to this conception. But what is *time?*

Time is a *mental construct* derived from the observation of duration. "Time" is not a thing. It does not exist as an entity of any kind, not even a *frame.* What exists are concrescences of *matter's energy* that, by perduring in existence display an ongoing continuity that we humans perceive as "time."[84] It reveals reality to be a *process* of the perdurance of *integrated functions. Time* is the human record of that *process,* measured by devices of our invention, and therefore a derivative of the human experience of self-identity.[85]

We started by saying identity is a function of time. Now we realize *time* is, more fundamentally, a function of identity

[*Author's note: My conception of time in this section differs from the way physicists Brian Greene and Sean Carroll imagined it. They said time is a function of entropy.*[86] *Entropy moves only in one direction, according to them, because the big bang was such a low entropy event, that the necessary subsequent movement to states of higher entropy creates the arrow of time.*

But from a slightly different perspective time is the **product of the conatus,** *the drive to exist, and it is the energy expenditure of the self-embrace of existence, that is responsible for the growth of entropy. Material energy creates continuing identity in only one direction because it's the only "direction" in which it does not yet exist. That is a metaphorical image. Actually there are no "directions" — direction doesn't exist until material energy creates the present "now" by existing, and the next, and the next, producing the arrow of time. Entropy results from the expenditure of* **existential energy** *which produces time. There is no such thing as "time" until material energy activates its drive to survive.* **It's the duration of existence that forges time** *where it does not yet exist. Time does not precede existence, it follows it ... and so does entropy. Time and entropy are the vapor trail of the soaring flight of existence, the glow from the furnace of creation.*

This understanding of time as a derived property of the energy to exist (and not an independent feature of the universe), corresponds to a similar understanding of

---

[84] Memory, of course, is the central instrument used for the perception of the passage of time. The human observer must, necessarily, remember herself and her *co-presence* with other entities in order to perceive those changes that accumulate to time. For without the perception of change, there is no experience of duration.

[85] Kant, *Critique:* Time is the "form of our inner sense, that is, of our intuition of ourselves and of our inner state." (A33, B49). Kant explored the significance of *time;* it became a key to philosophical thought thereafter. Hegel's "extrapolation" of Kant faithfully based itself on *time* by making *history* the locus of the elaborations of ideal reality, and Hegel dominated the entire 19th century.

[86] Brian Greene *The Fabric of the Cosmos,* NY Vintage, 2004, p.175. Cf also Sean Carroll, "The Cosmic Origins of Time's Arrow," *Scientific American,* Nov.2008, p 48ff.

space. *Space is not an "entity" in itself. It is the measurable perception of the relationship of matter's energy (as extended, as a non-point, as a collectivity of elements) within itself, to itself, and is therefore a derived property of the "internal" configurations of the substrate.*[87] *This way of looking at things concurs with the new understanding of* ***spacetime*** *that has emerged from the theories of relativity and the efforts to delineate a quantum gravity. Density and extension, the physical properties of the interrelationships within matter's energy, create, structure and configure space; and it is the structure of space (the bends and warps created by the densities of matter's energy or the "fields," the wave-like properties of particle relationships [gravitons]) that accounts for gravitation as a force that Newton erroneously believed defied materiality because it "operated at a distance." Gravity in the Einsteinian conception is a physical thing, the warps and bends in space, treating space as* ***something*** *that could be warped and bent.*[88]

This further emphasizes the unity of matter's energy as a global Whole. What we see when we look out on the Universe are not discrete, independently existing particles or their aggregates residing in an empty "clock-box" called spacetime. We are looking at an unbroken continuum, matter's energy extended and enduring, an organic Whole — ***one single continuous spacetime manifold*** *— a kind of energy flow, like a river, whose dynamic intra-relationships account for every last feature of our Universe as we know it, from spacetime to the complexity and intensity of our personal "selves."*][89]

This means that "becoming," as used here, is not only a term for discernible change; it means, in the first instance, a *sequence of temporal projections that maintain the* ***identity*** *of any stable modality of matter's energy* together with its matrix. We can call it a "series" of discrete "apparitions," or perceptions, but that would be an approximation. *Identity* is rather an unbroken continuum producing spacetime. Bergson, following Spinoza, calls it simply "duration." It is *the* primordial dynamism of *exis-*

---

[87] Cf. *Scientific American*, March 2005, Lineweaver and Davis, "Misconceptions about the Big Bang," p.36ff. The authors explain that the expanding universe is not "matter" moving under the impact of an explosion *into* a pre-existing space. *It is space itself that is being created by the expansion.* This can only make sense if we conceive of extension as a property of the expanding whole. Space is a derivative of matter's energy, a measure of its internal relationships, not something apart from it.

[88] In an hypothesis outlined in *Scientific American* Dec 2005, p.69ff, Jacobson and Parentani say that space behaves as if it were a kind of fluid, ironically evoking images of the old discarded theory of "ether." Light waves act like they were propagated in a *medium*, like every other kind of wave. Such a conception seems to allow spacetime to be understood as an elaboration of the inner structure of matter's energy.

[89] The evidence for quantum simultaneity over large distances implies either a communication faster than the speed of light (which is considered impossible) or a physical oneness of material energy that remains unexplained without jettisoning the "separate particle" imagery we have of matter. "... the entire universe has to be thought of as an unbroken whole." David Bohm, *Wholeness and the Implicate Order*, NY Routledge, 1980, p 222. (See appendix 5 here)

tence. Matter's energy, as *existence,* tends toward stable self-continuance (identity) through the establishment of relationships within itself, stemming from the integrative aggregations of the most primitive elements, quarks (vibrating strings of energy?). Self-continuance in time is identity ... that matter's energy *forms communities* of its own elemental constituents in order to perdure seems absolutely corollary. Communitarian integration appears to be essential to self-continuance ... and self continuance is identity.

In this foundational phenomenon, *matter's energy* is revealed as a "***congenital self-embrace.***" It constitutes all reality. At the highly complexified level of living organisms we call this dynamism of continuity, "survival." Spinoza, following in a long philosophical tradition, called it **conatus sese conservandi** — *the impulse to conserve oneself.* Identity is survival. But every coherent manifestation of *matter's energy*, in whatever form or combination no matter how primitive and undeveloped, even at the level of sub-atomic particles, displays the same characteristic dynamism: *matter's energy forms integrated functions, achieves stability and maintains identity; it extrudes a sequential ongoing continuum from which we generate the notion of time.*

This phenomenon, the self-embrace of existence, has been identified by many observers as the principal dynamism in the universe. Schopenhauer, as we have noted, called it *will.* Bergson called it *elan vital,* the Vital Impulse. Spinoza referred to it as a dynamic property of all things — **conatus** — that constitutes the very essence of reality. It is empirical, observable, in many cases measurable, in all instances, fundamental, universal and insuppressible.

We should emphasize: as Spinoza conceives it, this **conatus** is not a property or a quality of reality among and alongside others. It is, rather, **the very nature of existence itself — its inner energy.** All qualities, properties, activities, characteristics, aspirations, desires, joys, sadnesses etc., of whatever kind are derived from it. The **self-embrace of existence** is all there is ... it's all anything is ... there is no remainder. It is *material energy.* It's from this that identity arises.

In our analysis, *identity creates time.* It is the *persistence in existence*, moment after moment, and in the case of human beings, it is the sustained awareness of "selfness" through the sequences of changing psychic states which creates time. Time is an entirely *a posteriori* concept, a derivative of the human experience of continuing identity.

We experience this phenomenon interiorly. We call it "self." It is hardly exclusive to us, but the angle of vision it provides is extremely advantageous for our analysis. To experience what I'm talking about, take a minute. Get quiet ... and pay attention to your self as you pass through the ticking seconds of the clock. *You exist,* that is, you remain yourself, second after second ... but then again you are changing: having a sequence of thoughts, feelings, sensations ... you are a *continuum* of change, and the second hand of the clock is marking arbitrary divisions. This is *existence.* It is all there is*!*

two paradoxes of identity

Change. The first paradox of this connection between time and identity is that identity is not *stasis,* but *sequence. Time* reveals that **duration is dynamic**, not static. Identity is the sameness of an entity that endures precisely by *changing, i.e., moving beyond the "self" that exists* in any given moment and projects itself into the next and the next, and the next. *Existence is a dynamic self-elaboration* — a series of continuously consecutive epiphanies, if you will, that has no evident reason why it should ever end. It is a process of potentially unlimited self-extrusion, unlimited change. It is counter-intuitive to the static semantics, inherited from the Greeks, that we have heretofore used to describe reality. For it identifies the individual, the "one," immediately with multiplicity, the "many," in the sense of a changing "sequence of apparitions" which gives birth to our notion of time. It anchors the mystery of the one and the many in **process** and in **complexity.**

Identity, then, is *change, sequence.* It confirms that "being" is *becoming.* "Time" reflects the fact that the "essence" of reality *is process.* That means that *existence* is not "defined things" as the traditionalists suggest, nor random *"pointless"* interaction as the reductionists would have it,[90] but rather an *insistent enduring in time.*[91] **The energy of matter is a hunger for existence and necessarily moves toward maintaining it.** Reality is

---

[90] Random interaction is rightly identified as *unintelligible,* and I agree with those who say that recourse to it as an explanation is the abandonment of scientific responsibility. Cf Gilson, *Aristotle to Darwin* p. 130

[91] The term "pointless" validly refers to the fact that there is no *ulterior motive* to *existence.* The "point" is to exist, *to be-here.* It is far from meaningless.

not "something" *in process. Existential* process, *the extrusion of sustained presence,* is the only "something" there is. The "things" that this irrepressible process elaborates — the erstwhile "substances" of tradition — are the phenomenal display of the process of duration. At any given instant they are only the temporary platforms erected to sustain the process as it moves always to the next instant. "Things" are not the "goal" of the process; the goal is *simply to exist.* "Things" are instruments, tools, that serve the goal. **The only "goal" of existence is to exist**.

This means that *personality,* the human *integrated function,* is and remains itself a *process* equally focused on survival. My "person," is an evolving reality shaped by its interaction as an organism within a changing environment. *Time* is the measurable perception of this phenomenon — the result of my inner experience of the sameness of myself — my chronological "apparitions" in and by which I perceive that I am I, and determined to remain "I" within a changing context.

Structural ignorance. A second paradox, and one that is very significant for our analysis, is that **self-consciousness is experienced *separately from its biological infrastructure*.** We've already referred to this feature. The cells and hormones that make up my nervous and endocrine system provide a conscious identity for me that, *ironically,* is *totally oblivious to those very neurons' existence, character and role.* There is a firewall dividing the foundational apparatus of my bio-chemical organism from the conscious awareness of "self" it produces that is so impenetrable that I am forced to use a second pathway, external to myself, if I am to understand my own mechanism. I must use "science," not intuition, to *labor circuitously outside myself* in order to comprehend the very connections that produce my "self." No amount of subjective introspection can yield one iota of reliable knowledge about the biological structures or specialized functions responsible for self-consciousness.

We are conscious of ourselves and so we can see that we don't know what's going on under the surface of our own personalities. Other species also appear to be hungrily focused on their survival tasks and have no more awareness of what's going on within themselves than we do. The submergent integration of the tens of trillions of cells of my body, divided into specialized organs, is so complete that the substructures and functions of the organism are entirely absorbed in the **one experience of *self*.** They have *no independent self-awareness at all;* and even a consciousness as diaphanous as ours is completely unaware of their existence. This is the *integrated function*. As far as anyone can tell, it is characteristic of all conscious organisms.

personality

The blind dependence of the *integrated function* (the "self") on its organic foundation means that *personality is not a projection* or an illusion, much less due to a separable spirit. I experience my *self* as a transparent and hungry intentionality. But transparency and existential hunger do not imply immateriality, a disembodied soul that can exist apart from the energy substrate of the organism. My experience of self is as thoroughly organic as any other feature of biological life. We know it is a product of the human body even if at this moment we cannot give a detailed account of exactly what elements are operational. But we know they are critical to self consciousness because, as we noted earlier, if there is loss or damage to key locations in the neural network or the hormonal system, the sense of self disappears or is fatally altered.[92] *The material substrate is the dynamic organizing principle of living identity.*

Does this solve our problem? If we were determined beforehand that the answer *had* to be *either* a justification for the ancient claims of independent immaterial "spirit" on the one hand, *or* on the other, the vindication of the implications of reductionism (that the totality of being is *the only* individual, and individuals, which include the "person," therefore, is an illusion), we have definitely *not* solved the problem. What we have done rather is to reframe the question and generate an answer different from both. By eliminating both an imaginary "spirit" and an equally imaginary inert "matter," we have asserted the unity and integrity of the human organism (and by extension, every organism) and its intimate, genetic relationship to all organic life and matter. Any perceived duality of matter and spirit is attributable to our philosophical inheritance and to nothing else. There is no basis in reality for thinking there are two irreducible "realities" in the human organism.

Once the exclusively organic nature of human personality is established, two corollaries emerge. The **first** supports a perspective, like

---

[92] Cf Karl Zimmer, "The Neurobiology of the Self," *Scientific American,* Nov. 2005, pp. 93-101. In this regard, to claim that it is still possible to posit a "self" *able to exist and experience separated from the body* would require evidence. None of what has been offered so far is convincing.

Buddhism, that discourages the clamorous demands of the individual self for aggrandizement in the pursuit of an illusory permanence. There is no "personal" permanence; the only permanence is the indestructibility and infinite creative potential of *matter's energy*. The **second** connects the human individuals inextricably to their organic matrix, their forebears and genetically related contemporaries This means that the human individual is a part of a much wider all-embracing organic collectivity — a community of existential survival that reaches to the earth itself and beyond. This will have profound implications for our human social superstructures in economics, politics, and eco-morality. *The existential reality of any human individual, like any "thing," is inescapably subordinate to the Whole of existence.*

the idealist illusion

The Greek tradition which used the image of the craftsman (the "potter," in chapter 2) as a paradigm for the structures of reality, has prompted some observers to see this communitarian integration operating along the lines of human manufacture. A broom, for example, is a *material* multiplicity (straw, wood, string, etc.) organized around and by an embedded *idea with a purpose.* Here, the *idea* (to sweep) in the mind of the broom-maker is primary. The *idea with its purpose*, in other words, organizes the *material* and so creates the reality. It's like the *idea* in the mind of the potter, imposing form on the clay. *Mind and its ideas dominate and conform matter in these cases of human production.* This "insight" confused idealists like Bergson and Peirce (and Plato and Aquinas) into thinking that "matter" was a manifestation of "spirit" because they thought real organisms were like brooms or tumblers brought into existence by *ideas.*

But they are not. These items do not have the same inner relationship as the biological organism to its sense of self. In the case of the broom, the *idea with its purpose* comes first. In the case of organisms **the biological foundation comes first,** and the *idea*, if you will, the sense of self, comes second. While the *awareness* of my *self's* dependency on the organic substrate is reflexive, external, secondary and non-intuitive, *the dependency itself is not; it is primary and constitutive.* Early childhood development corroborates this view. The infant needs time to learn who she is. All personality theorists agree that human beings are born *without* *a sense of self.* [93] The vital impulse is not an independent conscious or pre-conscious "spirit;" it is *matter's energy* functioning in the cells which draw their energy and vitality from the molecules, atoms and ultimately particles that comprise them, and only later coalesce in self-consciousness.

The phenomena that correspond to the two perceived components of the human organism — the self and the substrate — are *the manifestations of one single integrated and developing concrescence of matter's energy*: this one organism.

summing up

We began this section 2 called "individuality — an interior view" with the idea of using personal experience to confirm and expand on the conclusions of our historical biological survey in section 1. Our conclusions are not insignificant. We have learned that the experience of identity — the interiorly recognized and urgent **self** — is a function of the perduring stability of the biological substrate which we know about only through science. The deceptive experience of a disembodied "self," we concluded, is created by the total submission of the cells of the body to the one individual. The "self" is not an insertion or imposition from without, nor is it an independent idealist "soul" that emanates our bodies, much less the product of our imagination. It is the conscious side of the integral coherence of the biological organism. It is *organismic integration* ... experienced.

It seems that our experience of what we call our "self," is simply an instant of the *existential* self-embrace, *the conatus* that we observed functioning throughout material evolution. It also seems reasonable to conclude that if such a *conatus* is found at all levels of integration, and manifests an unmistakable similarity wherever it is found, the phenomenon must derive from the energy of the most primitive elements. If that were not the case, it would require that some factor, *other than material energy*, responsible for those later manifestations, enter the process at some point along the way. But, *since there is nothing other than material energy in the universe*, not only would the operation of such a factor have to be proven, but the *very possibility* of its existence and provenance would

---

[93] Gordon Alport, *Becoming*, 1955, Yale U., pp 42 and 44. *"The young infant has ... in all probability no sense of self identity."*

have to be established. Besides, on the face of it, such an extraneous causal influence should be expected to manifest a *dissimilarity in behavior*, since it was dissimilar to the *material energy* it had begun to enliven. But by all evidence, the behavior is the same on both sides of the life divide; therefore ... etc.

We said earlier that cosmo-ontology went beyond science by looking at the behavior of *existence across the board*, i.e., at all the levels of e-mergence. ... So, what exactly does claiming to know what *matter's energy* **is,** its "nature," tell us, that science doesn't? Very simply, we have determined that there is a "living spontaneity" in *material energy* that manifests itself at different levels in different ways but is analogously the same in all things and cumulative. The *existential* energy at the most primitive level (proton formation at the first instant of the big bang) is repeated and activated in unexpected ways at each successive emergent level upward, not only in the form of more complex structures, but as a more far-ranging and more powerful *presence* — i.e., with new abilities to interact with its environment *and survive.* (And we speak appropriately of "upward" because these higher levels are achieved by including the lower levels and their functions even while going beyond them.)

This identifies the process that rules *all reality* as a hierarchialization. The perspective promoted here dares to interpret the energy on display *across* various levels of *existence* as *related,* and in fact, as *a homogeneous self-accumulation.* Material energy uses no other tools but itself, resident in structures it has already elaborated, to drew forth its new marvels. But despite the universal homogeneity, it has not, so far, been proposed as the subject matter of any science. I suggest it is the proper object of a philosophical instrument like *cosmo-ontology.* Cosmo-ontology studies *existence, material energy* across the disciplines and across the levels of emergence.

Material energy is the homogeneous substrate of all things. We all necessarily reactivate its fundamental dynamism and we are inescapably locked into its repeated strategies for survival. We are driven to do whatever this substrate is driven to do — *to survive,* which means *to exist*; and *as it exists,* with an irrepressible passion. Our range of activities is rigidly circumscribed by what *matter's energy* has been able to cobble together from negotiation between "things" and their environment through organismic evolution over billions of years.[94] We are, in other words, *THAT*. If the word "participation" has any validity at all, it must be applied to our genetic relationship to *matter's energy*. That *relationship* is where "we live and move and have our being." And if, besides all that, metaphysically speaking, we can also say *matter's energy* is an existential self-embrace, whose only "purpose" and intentionality is to exist,[95] ... an energy which is neither created nor destroyed and hence apparently cannot **not** exist ... might we dare describe it as "*esse in se subsistens*"?[96]

This contrasts sharply with dualist essentialism. Dualism assumes a bi-polar world of *inert matter* and *vital form*. It is a locked system. You can't change just one of those poles. "Matter" can be inert *only* if you have "spirit" to explain vitality. The world is indisputably full of life. Reductionists, you may note, are "dualists" without "spirit," because they treat matter as inert and offer no substitute for the role of "spirit." But we don't have to submit to the inconsistencies of a prejudicial reductionism. Given the undeniable evidence of vitality in the universe, *material energy* with its embrace of *existence* must itself be the source of life, because there is nothing else. *There is no separate spirit! ...* and our world is full of *life*.

Since everything that develops is constituted exclusively of *matter's energy,* shouldn't we expect to find the *integrated function* also operating at other levels of reality, both above and below the individual complex organism? Isn't this an area ripe for verification predicted by the hypothesis of living matter and the strategem of communitarian integration? Are there *integrated functions* at work throughout Gould's sixfold hierarchy of levels of evolutionary individuals, especially, as he claims, species?[97]

---

[94] This approximates what some have called Spinoza's "necessitarianism." But an examination of his moral counsels in the *Ethics* reveals that the "necessity" he is talking about are **conatus**-rooted "appetites" that derive from the organism itself. The "feelings" are necessary, but they can be controlled and their disruptive potential minimized, paradoxically, by understanding that they are, *as appetites*, uncontrollable.

[95] Aquinas, *SCG*, II, 35, 7

[96] I am reluctant to use the word "God" because it evokes the anthropomorphic deity of supernatural theism, **taken literally**. That "God" thinks, wills, plans, creates, acts in history, and providentially oversees all events like an all powerful human being. *Matter's energy* does not do any of that, except in the subsequent forms it evolves. The salient point is that if *matter's energy* **must exist,** as I claim, then participation in *existence* reinstates the Sacred but now as a universal possession. Traditional western imagery about "God" must be taken as a metaphor for *matter's energy;* "God" is a mythic symbol for *existence,* which is *matter's energy.*

[97] Cf Chapter 2,

3. individual and species, verifying the hypothesis?

C o-authors Bert Hölldobler and Edward O. Wilson are entomologists who specialize in the study of ants. They boldly state that *superorganism* is the only appropriate term for what they observe in the advanced social colonies of the genus *hymenoptera* which include ants bees and wasps.[98]

Recent studies, they claim, go beyond mere *analogy* to bring renewed interest in the **biological** similarities between complex individual organisms and colonies of insect and animal life, both in the way they are structured and in the evolutionary processes that bring them about. Their efforts intend to

> ... mesh information from developmental biology with that from the study of animal societies to uncover **general and exact principles of biological organization.** The key process at the level of the organism now seems to be morphogenesis, the steps by which cells change their shape and chemistry and move *en masse* to build the organism. The key process at the next level up is sociogenesis which consists of the steps by which individuals undergo changes in caste and behavior to build the society. The question of general interest for biology is **the similarities — the joint rules and algorithms — between morphogenesis and sociogenesis. To the extent that these common principles can be defined clearly, they bid fair to be recognized as the long sought laws of general biology.**[99]

Here we have a repeat of a familiar pattern. There are processes that not only exist but correspond to the same measurable biological mechanisms *at different levels of emergence.* We have proposed such phenomena to be the proper subject matter of cosmo-ontology because they indicate the presence of operators that transcend the focus of the sciences as they are currently conceived. By being common to many levels, such operators suggest that they reside in the common substrate, the *material energy* that forms both the organic structures and the environmental forces that drive them to evolve. It seems, as our hypothesis predicted, that *matter's energy* is a living dynamism that utilizes a *communitarian strategy* at all levels, confirming the homogeneity of the "nature" of *existence.*

---

[98] Bert Hölldobler and Edward O. Wilson, *Journey to the Ants*, Cambridge MA, Belknap, Harvard U Press 1994, p.110 *(emphasis mine).*
[99] ibid, p.111, emphasis mine.

It seems that in these examples of complex social integration, there actually exists a kind of species-level individuality, an inchoate "superorganism" that is not only a figure of speech, but a functional reality. A *superorganism* is an *integrated function* at the level of species or deme, capable of further integration in the future. Such integration has been discovered beyond the genus *hymenoptera* and is present to one degree or another among all animal species.

In their 2009 volume, significantly titled *The Superorganism,* the same authors make a more detailed, scientifically referenced effort to achieve what they proposed in 1994: *to find the common algorithms that rule both morphogenesis and sociogenesis.* They start out by stating clearly, "the ultimate agent of natural selection is always the environment."[100] And the response on the part of the organism explains the development of "superorganisms." The drive to survive has a collective dimension. Sociality emerges and is driven to ever greater complexity and division of labor, even reaching down into physiological differences between castes (in the case of ants), because the organisms that exist in such social arrangements absolutely outstrip any competitors in their ability to survive.[101]

Social insects help us understand the biological mechanisms in the evolution of sociality because, contrary to age-old popular belief based on the amazing complexity and flexibility of these societies, there *is no significant cognitive factor operating in the individuals' responses.* These creatures have brains the size of a grain of sand. The "algorithms" — programmed response sequences — embedded in those brains are necessarily few, simple and short. Hence they provide the opportunity to see evolution functioning in its "raw" form: the ability to develop complex interactive systems, whether an organism or a society, *in the absence of intelligence*, simply because they favor survival.

From our perspective, however, it supports the contention that the evolving characteristics of reality — like the *superorganism* — are neither introjected from "without" (as from an intelligent designer), nor the result of a self-transcending relatedness driven from "within" (as from an immanent "spirit" or seminal intelligence). Evolutionary communitarianism is the direct, simple elaboration of the very properties of *material energy* itself as it obeys its fundamental imperative: to survive, to be-here, *to exist by collec-*

---

[100] Bert Hölldobler and Edward O. Wilson, *The Superorganism,* NY W.W.Norton, 2009, p.28
[101] Ibid., p.29-30.

tivizing. It embraces itself. And complex communitarian structures are what it *always* ends up building, because they have proven themselves to be what works. Who decides that? **Natural selection.** How does this biological imperative play itself out at the human level?

human society

Alfred Kroeber, an early 20th century American anthropologist, taught that human society projected values that were not simply the sum of its members' in content, purpose and intensity. Society, he said, displayed an independent individuality as if it were itself a *superorganism.*[102] Steven Pinker has associated the great sociologist Émile Durkheim with the doctrine that the social group was *superorganic* and had a "personality" of its own, with feelings, goals, demands, values, commitments etc., *independent of the individual trajectories of its members.*[103] Was this a *metaphor*? Apparently Kroeber and Durkheim felt that, as scientists, they were describing a real phenomenon, not merely employing a creative teaching technique. In study after study Durkheim advanced arguments designed to show that the individuals of a society are, without their awareness, ruled and pressured into conformity by social imperatives that everyone had believed were indisputably individual. Suicide was the most striking example. One would have thought suicide to be the most individual, even *anti-social* of acts. Durkheim, however, said society directed *why and how* people killed themselves. He refers to *l'âme de la société* as if the human community were an individual entity.[104] It's as if society were using its members to construct a "higher" individual.

Sigmund Freud proposed that in the case of human beings, the purpose of "Eros," the "life-instinct" born at the moment of transition of inorganic matter to organic life, "is to combine single human individuals, and after that families, then races, peoples and nations, into one great unity, the unity of mankind."[105] There is no mention of intelligence or rationality.

But it's not only psycho-social theorists and cultural anthropologists that use that kind of terminology. **Biologists** universally make reference

[102] Steven Pinker, *The Blank Slate,* Viking, 2002, pp. 23.
[103] Émile Durkheim, *The Rules of Sociological Method,* (Paris, 1895), selection reproduced in *An Outline of Sociology,* ed. Robert Bierstedt, Modern Library, NY, pp. 305-314; cf. Pinker, *op.cit* pp.23-26; cf. Bernhard J.Stern *Historical Sociology,* Citadel Pr. NY, 1959, p.15
[104] Durkheim, *op.cit*
[105] *Civilization and its Discontents,* NY Norton, 1961 / 1929 p.69; he refers the reader to his own *Beyond the Pleasure Principle ... passim.*

to a "super-individuality" at the species-level when they speak of species or "nature" having this or that "purpose" behind certain adaptations. When questioned directly about such a manner of expression, they excuse it as *metaphorical.* And yet, the ease and directness with which they continue to make statements of this type (and their resistance to speaking other- wise) has given rise to the suspicion that they are, without acknowledging it, functioning under the influence of some form of crypto Lamarckism. Perhaps they harbor a secret belief that the species in question (or "na- ture" in a broader sense) is in some way *rationally directing* speciation, evolutionary development.[106] This practice of attributing *rational purpose* to "nature" is so pervasive in biology that Ernst Mayr, an uncompromising reductionist, one of the architects of the evolutionary "Modern Synthesis" of the 1940's, devotes two entire chapters in his *Toward a New Philoso- phy of Biology* to justifying it as a valid manner of expression despite its apparent contradiction to what he calls "orthodox" Darwinism.[107]

We are inclined to believe that biological terms however illuminating they seem, are only *descriptive metaphors* for sociological phenomena. Aristotle said that life is lived *only* at the level of the concrete individual

---

[106] Lamarckism refers to an evolutionary theory, proposed before Darwin, that attributed evolutio- nary development to "the inheritance of acquired characteristics." See Appendix for a discussion of a possible valid range for Lamarkism.

[107] Mayr, *op.cit,* "The Multiple Meanings of Teleological" p.38-66 and "The Concept of Finality" pp.233-253. Mayr argues that the final effect of adaptation is indistinguishable from intentionality and therefore that the teleological terminology is appropriate for descriptive purposes, so long as we understand what we mean.

My comment: An improper metaphor does not necessarily destroy scientific work, but it's hard to see how it doesn't *taint it.* The cosmologist or philosophical biologist whose task it is to understand the *significance* of the data that has been gathered, cannot function without protest in an atmos- phere where a principle of *organic biological function* like *species consciousness* or "nature's pur- pose" is routinely invoked, used universally in practice, then given perfunctory denial, and scientifi- cally correct explanations for concrete cases are not forthcoming. The "metaphorical" teleological explanation is often the only thing that makes sense, and scientists make little effort to show how such results could be arrived at *a posteriori.* If such "supra-organismic" rationality were to exist, it is a philosophical-biological fact of immense significance. It is dishonest and against the very principles guiding scientific enquiry to use the "species awareness" metaphor with such abandon, and then dismiss the significance that it implies without substantive justification.

The "selection" of sexuality as the favored means of reproduction, is a good example. Sexual reproduction is almost always explained as "nature providing the widest possible gene pool to in- sure the healthiest organisms." Despite its *universal* use, we must protest how "un-scientific" such a teleological explanation is. "Nature" provides nothing. Any such adaptation, by Darwinian stan- dards, is supposed to be unequivocally *a posteriori* — a result of what happened to survive, not an *a priori* plan for survival. But there is no serious attempt on the part of scientists to account for or correct these "para-psychic" attributions.

organism, the "substance." The *species* and its various sub-sets are derived notions that supposedly have *mental* existence only. It is clear that just because "humanity" is an *idea,* does *not* give it the right to be called a metaphysical *substance* in essentialist terms. But functionally, as our entomologists testify, the colonies of social insects provide material for natural selection. Group selection among the ants is a fact. Could human society, not unlike the various species of *Formicidae,* be some kind of *superorganism* impacted by the universal algorithms not only of biological but perhaps chemical and physical integration as well?

species level functions

We remember Stephen Jay Gould's lifelong efforts to prove that "species" are *also* the target of natural selection and that evolution goes on at levels above the individual organism. Gould's attempts at identifying "species as an individual" which we saw in chapter 2 were part of that project; it was a necessary component of his theory of "punctuated equilibrium." But, what form would material for selectivity take at the species level? Aren't all the genes that could possibly be the "carriers" and guarantors of species modifications for subsequent generations, locked into individual organisms?[108]

We all know there are functions, biological realities, *above the level of the individual organism* that bear on the ability of the species to survive that are organically incorporated into individuals, but are not "individual" traits. The "xx" and "xy" chromosome pattern, for example, insuring a 50-50 gender distribution in a population, *does not impact the fitness of the individual* in any way. Therefore it could not have been selected from among the random variations of an individual organism. Like so many other successful traits, it was the result of *group selection.* Obviously the survival of the individual organism was not the only factor functioning here ... or it might be better said that *the survival of the individual was a part of the survival of the group.* Hence selection worked on the group and the individuals with the group-favoring genes survived *as a result ... not as a cause.* As our entomologist experts say in *The Superorganism,* selection works at many levels simultaneously: the genes, organelles, cells, organ-

---

[108] All theorists, no matter how committed to random variation and natural selection, seem to require a fortuitous advantageous modification — a fluctuation or some other type of inheritable mutation at the level of the gene — for their theories to work. The "heritable mutation" is like *deus ex machina.* Cf *The Superorganism,* p.60

isms and superorganisms.[109] This suggests that survival is not exclusively an individual affair. The individual survives if the network (family, group, deme, species) insuring its survival survives.

divergence

Then, there is another phenomenon. It is an evolutionary process known as "divergence." Gould has a long discussion on divergence in *The Structure of Evolutionary Theory* and says it was identified by Darwin himself and ranked by him as *of equal importance with natural selection* as an explanation for evolution.[110] He quotes Darwin as saying divergence is "the tendency in organic beings descended from the same stock to diverge in character as they become modified." In *The Origin of Species* Darwin enunciates the principle in the following way: "... the greatest amount of life can be supported by the greatest diversification of structure."[111] Clearly, where competition is fiercest for the same nutritional niche, one can expect dominance by one or another extremely well-adapted species. But, in a fertile environment where there are many different food sources, to the extent that organisms differ — diverge — they will not be in competition for the same food source, and each will have a better chance at survival. Hence, on the *a posteriori side,* the amount of variation within any one species being supported by the environment as time goes on will become greater — exactly as great, in fact as the environment's different food sources will support. This environmentally stimulated variation will eventuate in new species.

Gould says divergence is a mysterious but empirically verified phenomenon whereby "life" (as it were) engages in "*superorganismal* (sic!) selection"[112] to maximize (one could say) the diversity among species within a given environment rather than promote (in a manner of speaking) the expansion of the most highly successful species. He suggests that such variation occurs *in order* (apparently) to take advantage of the multiple support niches existing there. As those words in parentheses indicate, it is difficult to describe the workings of this phenomenon without using terminology that alludes to *some kind of communication* between the environment and the species in question. Darwin himself in *The Origin of Spe-*

---

[109] Wilson, *op.cit.* p.24f.
[110] Stephen Jay Gould, op.cit, pp.224-250.
[111] Charles Darwin, *The Origin of Species*, 1859, reprint, (Random House), NY, 1979, p.157
[112] Gould, *op.cit.,. he uses the word "superorganismic" twice:* p.225, and p.227

cies began his chapter on Natural Selection (which contains his discussion of *divergence*) by saying, significantly, that "we have reason to believe, as stated in the first chapter, that a change in the conditions of life, by specially acting on the reproductive system, *causes or increases variability.*"[113] In chapter 1, under the heading of "Causes of Variability" Darwin alludes somewhat vaguely to changes in the "conditions of life" but makes one specific suggestion: "there is some probability in the view that this variability may be partly connected with *excess of food.*" He speaks of the new conditions of life "*causing variation*" (using those exact words) in the organisms "exposed" to them.[114]

Such "*causation*" might be understood to imply a primitive recognition of and response to the environment on the part of the species in question — *a communication* — making variation occur (lifting genetic constraints) not just randomly but *in the direction of exploiting the available food source*. "Selection" then has what Darwin calls a "profitable" variation to work on to develop new species.[115] This would operate in tandem with standard evolutionary principles. The thrust to variation, retarded by limited survival supports, under conditions of abundance will produce varied offspring that are more apt to survive and reproduce. But we cannot fail to see that there must be some kind of *communication* between organism and environment for this to happen.

Divergence means that "life" tends to "fill out" whatever environmental openings are accessible, as a tree will fill out with its leaves and branches all the regions of available sunlight. In the case of the tree, there is no identifiable center of consciousness or intelligence, like a "brain," and yet we have no trouble attributing "perceptiveness" to the primitive sensibility of the arboreal individual "recognizing" sunlight and actively responding to it. Why should such a phenomenon be considered completely unimaginable in the case of species sensing and responding to a survival factor of such importance as a food source?

We know there is perceptiveness in the individual organisms which constitute species. Even more than a tree, an animal can "sense" the availability of food in a niche for which she, perhaps, was not originally adapted. Exploiting the new food source would introduce a behavioral

---

[113] Darwin, *op cit*, p.131
[114] Ibid., p.71
[115] Ibid., p.132

modification based on *learning*. Wouldn't those of her offspring, *taught* by her, who happened to be better at the new food gathering "trick" because of some structural or functional variation peculiar to the individual, be selected? After all, if she were already structurally capable of the "trick," it's unlikely her progeny would not be similarly endowed. Selection is still in the driver's seat; but it works together with *learning*. Learning also affects the choice of mates, as those who can exploit the new food source will congregate and live long enough to have reproductive success with one another. This would guarantee that subsequent generations would continue working in that new niche. Wouldn't selection then favor those *incrementally accumulated and increasingly effective variations* in the same direction until the organism was thoroughly adapted to the new source of food?[116] The result will be organismic changes tending toward divergent speciation ultimately directed by a variability which, *while it was not entirely due to learning, we have to admit was not entirely free of it either.* Darwin himself thought "natural selection" was secondary here, for he said, "... unless *profitable* variations do occur, natural selection can do nothing."[117] The origin of species, then, would come down to the origin of the "profitable variation." This might suggest a modified Lamarckism, or a nuanced Darwinism.[118]

interaction with the environment

This type of variation would be an interactive phenomenon that, because of *communication and learning*, by which I mean a response to available *information,* transcends the random reactions of the reductionists' imaginary "inert" matter.

> [**Author's note:** *the words "communication" and "learning," in this context, like "information," "intentionality," and "life," is applied analogously (metaphorically) across the phyla to refer to interaction with the environment that involves the organism's* **autonomous ability to redirect its activity in response to stimuli.** *Sometimes it's a metaphor — depending on the level of function. An organism's activity is not a mechanical reflex. But by the same token it is certainly not valid to think of it as a conscious operation, which the words "communication," "learning" or "information" would imply, if taken literally.*]

Is it entirely impossible that organismic individuals might carry a "sense of the species," an instinctive, subconscious awareness of the species'

---

[116] the example was used by Dawkins *The Ancestor's Tale*, p.199
[117] Ibid., (emphasisi mine).
[118] See appendix on Lamarckism.

collective biological needs not unlike the sensibilities of the tree? A species does not have a central "brain," but it may carry out its collective functions through the "species-sensitive brains" of the individual organisms that are its members, as trees do through their leaves. Such a collective instinct embedded in individuals would be consistent with the theory promoted in these reflections: that the transparency with which the dynamism of *matter's energy* embraces itself, tends to produce *at all levels* (above and below the individual organism) *integrated functions* — rudimentary inchoate individualities — **communitarian identities** with a sense of "self" gathered from its "cellular units" and a corresponding investment in *survival.* This would include the possibility of a "will" or intentionality capable of nudging evolutionary development in certain directions rather than others. What I'm suggesting has nothing supernatural or "spiritual" or "intelligent" about it; *it does not imply conscious purpose or rationality of any kind.* I'm talking about the survival drive. It is that same propensity of matter's energy, analogous to what we encountered in nucleosynthesis, in the development of complex molecules, multi-cellular organisms and the self-conscious human personality, whereby material indentities persist. They survive. It's the self-embrace of *existence.*

I believe it is entirely tenable, once we realize that **the tendency to integrate is an inherent property of material energy,** to expect that *over time* more and more complex — more integrated — organisms will evolve, just as they have in the past. Once again there is nothing here that requires that it be explained by a chosen "plan" or outside intervention. We are just carefully observing *what matter's energy has always done.* There is an unmistakable pattern here: integrated functions, under the pressure to survive, have issued in exponentially more complex integration, even though it was impossible predict in advance what form or direction they would take.

emergence: new structure from integration

The highly complex evolutionary development which we as humans are, was built upon more primitive life forms, going all the way back to the original prokaryotic unicellular ancestor of us all, as we've seen. The astonishing fact is that we *emerged,* eventually, from something that we were not. "Emergence," in evolutionary biology, is a technical term that has a specific meaning:

Emergence, in evolutionary theory, means the rise of a system that cannot be predicted or explained from antecedent conditions. ... The evolutionary account of life is a continuous history marked by stages at which fundamentally new forms have appeared: (1) the origin of life; (2) the origin of nucleus-bearing protozoa [eukaryotes]; (3) the origin of sexually reproducing forms, with an individual destiny lacking in cells that reproduce by fission; (4) the rise of sentient animals, with nervous systems and protobrains; and (5) the appearance of cogitative animals, namely humans. Each of these new modes of life, though grounded in the physico-chemical and biochemical conditions of the previous and simpler stage, is intelligible only in terms of its own ordering principle. These are thus cases of emergence. [119]

This speaks to a *potential* and even a *proclivity* within matter's energy itself for "integrative participation" — the shared organic configurations of living entities as they lend themselves to the formation of emergent collectivities that arrive at new levels of functioning for survival, *not predictable or even anticipated* by the previous levels. These are the properties of *matter's energy* as displayed in the increasingly complex life and social forms that have *emerged* in evolutionary history. Notwithstanding our inability to predict what will emerge, we would have to admit that, because emergence has occurred so many times in the past, it seems likely that such advances will occur in the future. Emergence shows what *matter's energy* has done and is capable of.

Integration is selected

A bee hive is an instance of the *integrated function* operating at a level above the individual. It is a "superorganism." Bees are of the genus *hymenoptera,* and we may have assumed that the way they conform their societies is the result of some "hard wired" feature of the entire genus. But the remarkable thing is, there are many more species in that same genus that are communal on a much less intense basis, like bumble bees, carpenter bees and andrenids. This suggests the *superorganismic* lifestyle is the evolutionary *development of a potential* — it was a variation that was *selected*. If it was selected in the case of one species, it can be selected again, given the proper conditions, in another species *because the communitarian potential is there in every element and force of material energy*. In fact, sociality in one form or another has been selected in virtually every species known to science. Species (or deme) selection is a

---

[119] **"emergence."** *Encyclopædia Britannica* 2006 Ultimate Reference Suite DVD. See Appendix 3 for further readings on emergence from the *Encyclopaedia Britanica*.

fact; its presence indicates the existence of the communitarian potential operating throughout the phyla of living things. .

At one point in time the unicellular individual was the only life form on earth. But it was individual in appearance only, for within itself it was an immense collectivity of complex molecules gathered and forged from the environment and pressed into living service to the cell. The subsequent development of multicellular organisms from colonies of those unicellular individuals was due to *matter-energy's* natural tendency to continue to integrate. There are examples of this transition in the truly remarkable *social microbes*, one of which, the *slime mould,* a social amoeba, is described in detail by Richard Dawkins in his book on evolution, *The Ancestor's Tale.* Dawkins says it "literally blurs the distinction between a social group of individuals and a single multicellular individual."[120] He describes how the independent and individual amoebas

> ... rather abruptly switch into 'social mode.' They converge on aggregation centres from which chemical attractants radiate outwards. ... Eventually the amoebas unite their bodies to form a single multicellular mass, which then elongates into a multicellular 'slug'. ... The amoebas have suppressed their individuality to forge a whole organism.[121]

Dawkins then describes how the "whole organism" functions as a reproductive fruiting body to spawn new, individual, independent amoebas, beginning the life-cycle all over again.[122] It seems the tendency to integrate, aggregate and complexify exists wherever *matter's energy* is found.

At another level, even further down, there is the example that we have already used in this chapter. Our own individual cells, known as **eukaryotes** (with nuclei), display evidence of being the resultant form of a more ancient multicellular community that had incorporated **prokaryote** (more primitive, non-nucleic) bacteria-like cells functioning as metabolic and chemical processors within the cytoplasm of the host cell. To this day these "cellules" (mitochondria and chloroplasts) exhibit their own peculiar structure, metabolism and reproductive mechanisms that are different from those of the eukaryotic host.[123] So even our individual cells are them-

---

[120] Dawkins, op.cit. p.504
[121] Ibid. 504
[122] Ibid. p.505
[123] Goodsell, *op.cit.*, pp.70-72

selves the *integrated functions* of aggregations of more primitive living entities.

conclusion

All this serves to confirm the view that the tendency toward the *integrated function* exists at all levels. It is not a conscious, "spiritual" or supernatural phenomenon; it is not the result of a "plan" or a provident intervening rational "God" nor exclusive to *rational animals.* It is operative throughout the realms of the material universe proportionate to the level of complexity achieved. We conclude that it arises from a potential of *matter's energy* itself, an *existential* energy that is a "congenital self-embrace." It makes *stable, survival-orientated aggregated integrations* available for selection whenever conditions favor such adaptations. It is a prime strategy of survival.

We can say as a minimum, therefore, that *matter's energy* seems capable of producing entities that are stacked and nested, reflecting a multi-level confluence of identities — a communitarian *"participation"* — that is *not* the result of neo-platonic conceptual predication or taxonomic classification, *nor* is it an escape into an imaginary immaterial spiritism. *It appears to be a physical potential for cooperation, integration and communitarian co-existence embedded in the very energy at the core of existence itself.*

This "participation," unlike Plato's eternal fixed ideas, is material and dynamic; it is reproductive and therefore, broadly speaking, genetic.[124] It produces family. All things are the members of a single community, the extensions of a single dynamism which coalesced into baryons at the very inception of its trajectory in the big bang, and has displayed the same patterns repeatedly thereafter. *Matter's energy* functions beyond the random billiard-ball extrinsic effects once believed by a mechanistic reductionism to be the only characteristic of material entities. The **conatus** in other words, is a feature of *matter's energy* that, despite the undeniable empirical evidence, is not included in the reductionists' inert view of matter. *The reductionists are wrong!* They cling to a flat, one-dimensional notion of "matter," a hold-over from a prejudicial rationalism — a mechanistic Car-

---

[124] **"genetic":** here I am not speaking *only* about the genetic transfer of traits that takes place through the processes involving DNA & RNA which is the usual meaning of the term, but also the fact that even the distinct innovations that are unmistakably *emergent,* have inherited everything that they are from the forms from which they emerged.

tesian variation of Platonic dualism, born in the the 17th century when the existence of "spirit" was an incontestable assumption, the correlate of an inert "matter." They would reduce all activity in the universe — chemical, biological, organismic, conscious, societal — to the interactions observed by physics.

Reductionism treats "matter" as if it were a hollow empty shell, utterly flaccid before the whims of "mind."[125] But matter is not dead and inert; it is energy for *existence*, it is an irrepressible drive, and in its most complex recombinations, like humankind, is virtually beyond control. Mind is matter's derivative, not its source and director. But even without a director, the *process* is not directionless; it has a bearing, a direction, an intentionality. At the most fundamental level, and therefore at all levels and all times, wherever *matter's energy* is found, **it displays an orientation toward existential persistence**, a non-conscious or pre-conscious self embrace — and that has always meant communitarian integration. The very *process* which is *existence* issues in multi-faceted and multi-level entities, communities of individuals, which we ourselves are privileged to experience interiorly as *our very selves*. Each complex organism is such a multi-level community. In fact, we may be experiencing the beginnings of our individual incorporation in another, "higher," social integration. These are the processes active in evolution. We have come from them. We are part of them; we do not create them nor do we control them. They are an abiding potential because they reveal the natural properties of *matter's energy.*

Can we go further? We know the survival drive is there and functions. But why? Why must matter's energy "survive" in structured form, perdure, and toward that end, aggregate, collectivize, replicate, complexify and evolve, leaving in its trail time, space and a rapturously beautiful universe with at least one planetary world abounding with life? And must *existence* build communities and therefore "higher" individuals out of other individuals ... one thing out of many? Is there no other way to survive? Was Teilhard right — does it portend the future of humanity as a *superorganism*,

---

[125] Etienne Gilson, *From Aristotle to Darwin ...* tr. John Lyon, U of Notre Dame, 1984, p.19 and *passim*. Gilson accuses Descartes of promoting a mechanistic cosmology specifically designed to give humankind absolute control over material reality. He quotes Descartes' definition of the object of philosophy: "... [to] *thus render ourselves the masters and possessors of nature."* (René Descartes, *Discourse on Method*). This confirms my contention that the inertness of matter is a *dualist* projection ... it is a corollary of the belief in the existence of vital "spirit."

an integrated function of cosmic proportions? Are we, as he suggested, going somewhere? What would such a construction mean for us?
   We deal with that question in the next chapter.

chapter 4
eschatology

existence and eschatology

S ome contemporary evolutionists are persuaded that with the sub-strate's ability to form *integrated functions* we are looking at the source of a structured macro-evolution. They claim that Evolution *necessarily* implies increasing complexification through time. Some even believe they see a *plan* behind the upward spiral of species. Pierre Teilhard de Chardin and Alfred North Whitehead are well-known modern proponents of this vision. C.S. Peirce, acknowledged as the greatest American logician and philosopher of the 19th century, projected that the world would become "an absolutely perfect, rational and symmetrical system, in which mind is at last crystallized in the infinitely distant future."[126] They all claim to discern a communitarian future, an *integrated spiritual function* of divine proportions for humankind at the term of the process. De Chardin, for his part, believes the path was pre-ordained by "God."

The belief in a *pleroma*, a completion and divine fulfillment to the process of *existence,* is nothing new in our tradition. We in our era are not the first to recognize the central importance of *process* in the character of reality, and it seems that once you speak of *process*, the question of goal and direction arises naturally: ... process means things are *going* ... doesn't that necessarily mean they are going *somewhere*? Is the process we see just random movement or is it really a *growth*, leading to a fixed (and possibly planned and desired) end — what the Greeks called an *eschaton*? Here I would like briefly to reflect on some of the thinking — the *eschatology* — of process philosophies in our tradition. Science, by the way, has no opinion on this matter; and, from an orthodox reductionist point of view, scientism denies all teleology of whatever kind.

---

[126] from "The Architecture of Theories," Weiner. *Values* ... op. cit. p.159

rational plans?

First, we should observe that expecting a fulfillment implies that the *process of existence* has a *purpose* ... a purpose beyond itself. It is characteristic that these philosophers, in contrast to the proponents of scientism, see a *rational plan* behind the developments of evolution. Such assumptions are not surprising because, as we've seen, the tradition we've inherited has been thoroughly *essentialist, idealist* — ruled by a metaphysics of *ideas*, ... and ideas are the products of *minds* that make rational plans with **intention and purpose.**

As Spinoza enunciated so clearly in the Appendix to Part One of his *Ethics*, human beings are focused on "final causality," what we call "purpose." *Purpose* is the ruling category in human intelligence, and a proof that the entire conscious apparatus evolved as a function of survival. For the introduction of *purpose* into a "pointless" universe is entirely gratuitous; it does not correspond to the "nature" of *existence*. It is rather the invention and intervention of the **conatus** which must achieve definitive conquests over dissolution — the securing of food and shelter, the avoidance of hostile forces, etc. **Conatus** has to make a "point" of providing survival, for the unmodified environment cannot be relied upon to do so. It seems hardly necessary to make the additional point that human "rationality" would most naturally be projected onto a "God" that has been conceived as a "person" like us. A "God," a Master-*Mind* with a rational plan, has been the fundamental *essentialist* scaffolding of the Greco-western view of things, giving structure and intelligibility to the otherwise incomprehensible "pointless" welter of existential process events. But even by scholastic standards **"God" is not "rational"** because he has no "reason" beyond his own existence *for doing or not doing anything!* But since "he" is **esse in se subsistens,** "he" has all the existence there is and so needs to achieve nothing. Please observe: like *matter's energy,* the Thomist "God's" only "goal" is *to exist.*

Secondly — and derivative of that — expecting an end-product implicitly asserts that the *process itself cannot be the ultimate reality* because it is subordinated to something else, something beyond itself toward which it is going. If the *end product* is the *reason for which* the process exists, then the end-product — that "something else" — would be the "real reality." It would be the desired goal toward which the process is ordered. The process in that case would be secondary — subordinated to it as means to end. Furthermore, the *process* is also subordinated to the *plan* and the *planner* (if there were a "plan," there would have to be a "planner" ... by definition), guiding it at least, and probably its Creator. But we know that is *not the case! We can see from the way it behaves, existence has no goal beyond itself.*

If there's a **term** or finality, then clearly, there would have to be more to *existence* than *process*. This would be contrary to what we have been discovering in these reflections. We have looked at *existence,* and saw that whatever we can observe of it, always and everywhere, is nothing but *process.* The "nature" of existence, we concluded, is *process* — the time-generating process of staying here.

An end product, if it were *not* just another *process*, would have to be something fixed, a state or a "thing" — complex, perhaps, but neverthe-less static and fixed and finished ... or else it would have to be *another process.* If, indeed, it *were* another process, it would have to call for another term or end product under the same conditions as the first, again, either something fixed or another process. It seems clear that if the "goal" of the *process* is itself a *process*, the word "term" does not really apply. For a "term" that issues in another process terminates nothing.

So it seems we cannot define *existence* to be *process* without eschew-ing a final term. Once we decide that *existence* is *only process,* there can be nothing other than *process* (even if that means that the *process* takes a new form). There is no fixed term ... ever.

shocking?

The fact that we find ourselves somewhat shocked by that statement reveals how deeply embedded the traditional eschatology is in our minds. A universe of *endless process* is counter-intuitive ... it feels unsatisfying. And yet, however exhausting it may seem, at least we can imagine it. Endless is simply the continuation of *existence* in sequential time as we know and experience it everyday in our universe. We are familiar with *existence-in-time.* And, the only alternative to an endless process is a fixed unchanging state or "thing" that lasts forever — something eternal. While that outcome feels more intellectually satisfying, for it corresponds to our human survival drive which functions through intelligent purpose, planned goals, fixed ends to problem solving and a "snap-shot" conceptualization[127] apparatus, we have to admit it is *absolutely unimaginable. We can imagine endless, but we cannot imagine eternal* for we have never experienced anything like complete *stasis*, utter completion, eternal changeless perfection. Our experience is limited to *process.* Pure *eternal stasis* is a projection — the product of our imagination dominated by the archetype of purpose.

This does not deny "eternal" as a possibility, but it emphasizes that it is not derived from any known experience. It is not an *a posteriori* concept. It is an abstract projection, not unlike "being" and its correlative, *nothingness*, of which it is corollary. These are all products of our imagination.

eschatology, ancient and modern

In the "modern" views of de Chardin and Whitehead, the integrating tendencies of *matter's energy* presage the emergence of a *consequent* "Divine being." Theirs is a variation on the traditional theme established in ancient times. The ancient neo-Platonic philosophers like Plotinus and Proclus believed that the universe is a *process* set in motion by a divine being and that the *finite existence* of each entity is essentially a needy dynamism that seeks fulfillment by returning to its Source. The word used for this fulfillment was *"pleroma"* — a divine *"fullness"* at the end of time that would *fill up* the need inherent in finite "things." A Christian eschatology built along these general lines is found in the earliest NT writings.[128]

The ancient Greek vision proposed that all things result from the emanation of *spirit* out from a *primordial source, a Pure Spirit* they called "The One." The One was Being Itself, creating beings each with a degree of "being" (= *spirit* = mind) proportionate to their *essence* which also determined their *purpose* and end. The metaphor they used was a light source, like the sun, radiating *spiritual* light out into the surrounding *material* darkness. The light was *spirit* which is being; the darkness was *matter*, nonbeing. The closer you were to the sun, the more spirit you had, the further away the less spirit until the rays of light petered out in the total darkness of formless matter, nothingness.[129] Pure Spirit was Being and Light and Eternal Life, and it was the radiance of that Spirit that created life, nature's

---

[127] see chapter 5 for the elaboration of this notion.
[128] Cf Rm 8:18-23. This antedates Plotinus by 200 years.
[129] The "light" motif was being used by Christianity as early as the gospel of "John" (early 2nd c. CE).

dynamism, as an echo of its own. Being's radiance thus set in motion a *process* by which all things sought Eternal Life by returning and reintegrating with the One, the Pure Spirit from which their spirits had arisen. Since created spirit was now mingled with "matter" (*non-being*, "darkness," the carrier of the contagion of death), the return to the One which achieved immortality *could only be accomplished by shedding "matter."*[130] The process, therefore, was clearly in service to another end — *the pleroma* — an end totally dissimilar and far superior to the process designed to produce it. At the end, the process of shedding matter would no longer be needed. There would be no more process, because there would be no more matter. The need to "overcome" matter explained *becoming*. Once spirit was able to accomplish this, it automatically became eternal and unchanging. Eternal unchanging spiritual immortality was the goal.[131]

the schoolmen

In the middle ages, Aquinas also imagined a creation scenario that was focused on *process*. In this view, "God's" overwhelming prior timeless and immaterial presence, *without any activity on "God's" part whatsoever*, who simply "wills himself," draws all things into spatio-temporal existence "out of nothing" by the sheer force of irresistible attraction. It's as if "non-being" could not resist the attractive power of *Being*. Existential development of any kind, in other words, would be seen as the delirious upward attempt of "non-being" to "become" (by participative imitation) the Divine "Being" of "God," thus spawning finite spacetime. *Time*, in other words, is simply the duration of the *process* set in motion by the Divine Presence. The "end" would be the natural term of this "creation process." A modern process thinker like de Chardin could fit quite comfortably within this scheme of things. He might only add that the *entire creation process* imagined by Aquinas is what we now know as evolution.

We should keep in mind that the "priority" which mediaeval and ancient theory assigned to "God," is *not* temporal and physical but rather purely (and I would add, unimaginably) *metaphysical*.[132] That means it describes an **ontological relationship** between Creator and creatures that is *com-*

---

[130] Paul's projection in Romans 8:21 that all creation "will be set free from its bondage to decay" implies a sloughing off of the impediments of matter, if not matter itself.

[131] Plotinus, *Enneads*, I tr 6, nos. 2,3,4,5

[132] Aquinas, *SCG*, II, c 18,2

pletely independent of any physical contact or the sequences of time.[133] The Aristotelian-Thomistic concept of final causality exemplifies this unimaginable relationship perfectly. An immaterial "God" is imagined to be the metaphysical source of the material beings spawned by the generative power of the Divine Presence alone. There is no Divine activity in time or space and so there is no before or after for "God." "God" is Pure Spirit, completely immaterial, completely impassive, and there is no possible "physical" contact with creatures. Creation is not a work *ad extram,* in the Thomistic scheme, it is *internal* to "God."[134] It is, in fact, an aspect of "God's" self-contemplation, and identical with the divine essence itself as are all "God's" "actions" and attributes. "God's" eternal bliss is to contemplate (and "will") "Him"-self. Seen from the perspective of a "self-transcending" generosity, the schoolmen would say, it is called "Creation."

We can discern the radical *idealism* at the core of this vision because it asserts that "God" is *Thought,* and creation is "God's" *thought,* a self-image *by participatory imitation* and therefore *existence* is, ultimately, *ideas,* implying a master plan. Such a plan, moreover, is *rational,* because it has a *purpose* and we humans can understand it at its root — for we *know essences* and their purposes. The idealist theories advanced in the 19th century by Hegel and his followers simply took the traditional vision a step further and added the dimension of *time.* Not only material creation, they said, but also human history, our own living process, both individual and collective, was the elaboration of "God's" inner spiritual reality. It was divine *thought* manifest in time. *Idea,* the divine plan, purpose, dominates the process.

matter and spacetime

Why the universe should be material — and reside in spacetime — and not purely immaterial and dwell in an eternal "now" of *Thought* as "God" does, is not explained in these theories.[135] Nor are we given to understand how creation can be ontologically constituted by an involuntary *process-toward-God* without making human choice irrelevant, and a halting, geologically extenuated evolution mere theatrical appearance.

---

[133] Aquinas, *SCG,* II, c 17; c 19
[134] Aquinas, *SCG,* II, 23,5; *ST,* I, q.28, a.4
[135] Actually Hegel and other **monistic** *idealists* like Peirce were more consistent in this regard. For them, matter is *only phenomenal,* i.e., it is *only appearance,* since "being," metaphysically, is *only thought, idea.* This means that matter is not an inert something different from and opposed to spirit. They are one and the same thing.

In a dualist universe, "matter" and its associated emanation, space-time, is an anomaly, if not a contradiction, and must be explained, in practice often by strange paradoxes like calling matter "non-being." These *necessary* equations — of matter to non-being and spirit (ideas, mind) to being — created proportionalities that were later reflected in the Thomist doctrine of the *analogy of being*. Bernard Lonergan, who is a Thomist, calls metaphysics "the integral heuristic structure of *proportionate* being,"[136] But to call being "proportionate" is to accede to the *idealist* premises granting ultimate reality to *ideas*, spiritual thought, and mind, — more mind = more spirit = more being — and impute *non-being* to matter. It established humankind at the pinnacle of earthly existence. But it was a lonely and alienated Valhalla: Despite the Neo-Platonic vision of a graduated distribution of "spirit" among emergent species, in the hands of "christian" interpreters we humans were left the only spirit in a world of matter.

I believe the traditional Thomistic view, elegant and all-embracing though it may appear, is as obsolete as the matter-spirit dualism from which it was derived. *Existence* is not divided into matter and spirit, and matter is not a negative, death-bearing "non-being." In the Thomist era, participation was still determined by the classifications created by word-labeled concepts, and a metaphysics based on *essences* and an imaginary world of spirit.

Aquinas believed he was working from absolutely unassailable premises established by "revealed facts" (guaranteed by the Church), *viz.*, the existence of "God" as "Spirit," which was taken to mean, in Aristotle's terms, that "God" was Pure Mind, immaterial, absolutely static, immutable and impassive, and therefore the Creator-creature relationship was "purely metaphysical" as we've said. It confirmed the essentialist interpretation of the "concept of being" as infinite and necessary, instead of finite and conditioned (as we actually find it). Without those premises the ancient / mediaeval vision has no ground to stand on. It is pure projection. Our own definition of being-*as-existence*, coming from the experience of concrete *presence-in-time*, does not recognize a gratuitous creating and sustaining relationship to the physical world from something *outside* of it ("God" an immaterial spirit, eternally static and unchanging, creating matter and a world of time and change).[137] I deny the existence of "spirit" as a

---

[136] *Insight*, p.391
[137] Gregory of Nyssa, *op.cit.*, XXIII, 3 and 4, expressed it well. See appendix for text.

separate category of being. Therefore I also reject any immaterial "onto-logical" relationship evidenced *only* in the logical relationships among concepts — *ideas* — accessible *only* to the supposed immaterial workings of immaterial minds. If "God" made the world, or it's substrate *material energy,* "God" must *be* what the world *is,* and must relate to the world according to the mode of existence of the world — as *matter's energy,* a reality in *process.*

The original error of the ancients, I believe, though it was hardly avoidable given the myths they inherited, was the inclusion of the category of "spirit" in the "concept of being," a corollary of the divinization of *ideas,* allowing for the scholastics' projected *metaphysical* relationship and fixed and static definition of reality. But **we do not encounter existence in a static and fixed form anywhere.** And all *process,* by essentialist standards, is necessarily finite and conditioned because it "lacks" what it is becoming. The only source for the mediaeval conception of an immutable immaterial "God" *is our imagination.* Since I do not accept the imaginary division of reality into matter and spirit, I cannot accept the "metaphysical" derivatives and implications of that dualism. "Being," I believe, is **only one thing**: *existence, matter's energy, a self-embracing process that creates time.* Therefore if "God" *exists,* "God" must be, or must emerge from, *material energy,* and must be related to the world *as matter's energy* which is naturally and endlessly in *process. "God," in other words, must be integral to the becoming of the universe* **such as it is.** The alternative is to say that there are two diametrically opposed **existences**: the one a finite and temporal material process, and the other infinite, immaterial and unchanging and the first somehow **becomes or morphs into** the second. Hardly even thinkable much less compelling. No, as Spinoza insisted, "God is an extended thing."[138] And if that is true, then we can experience "God" ... and through that experience, we can *understand* "God," for what "God" is — *material energy* — is what we are.

The goal of these reflections, after all, is *to understand.*

the modern millenarists

Both de Chardin and Whitehead, working from the data of modern science, recognize the autonomous creative power of matter in the evolution of species ... and try to explain it. Whitehead sees matter as the *locus* of the divine presence itself and so he unabashedly calls *matter's energy*

---

138 Baruch Spinoza, *Ethics,* II, p.2

the "Primordial Nature of God." De Chardin, more invested in avoiding pantheism, believes a Creator "God" placed an "interiority" in matter as the seed of consciousness. But for both, despite their acceptance of the "reality" of matter and the dynamism of material process, their distance from the traditional conception is not that great.

Consider: Whitehead and de Chardin each require an essentialist element (ideas) embedded in matter for their visions to work.[139] They project a physical and biological participation throughout development, producing a *recognizable* but fixed and finished "divinity" at the *temporal end*, the Omega Point, based on the process we observe unfolding in evolution.[140] But in our reflections we have learned that all the forms that *existence* has taken, all the species, all the achievements of evolution, are *integrated functions*. They are the constructions of unplanned groping, discoveries, the *a posteriori* outcome of the evolutionary drive to survive achieved through *functional integration* interacting with the conditions *existence* encounters in the real world. *There is no plan!* Phylogenesis, the creation of species — evolution — is adventitious. It just *happened to happen* that way, driven by the paroxysmal embrace of *existence* in an ambiguous environment to which it had to adapt or be exterminated. *There is no plan!* The process is focused only on *existence*.

In the views of both these men, the apex of development is an emergent divine *entity*. But we would have to hypothesize that any entity that emerged from *existence* as we know it, would most likely be an *integrated function*, because that's the way *matter's energy* works, always and every-

---

[139] In Teilhard's scenario *matter's energy* is a Sacred "potential" from which everything physically draws its "upward" evolutionary thrust. He identifies this potential with a directedness he calls the "within" of matter, its interiority. This "within" he unapologetically defines as consciousness (*Phenomenon of Man*, tr. Bernard Wall, Harper and Row, 1975 (1955), p.57, and 53-74 *passim*) gratuitously ignoring its origin in the "drive to survive." Whitehead, for his part, also requires that primitive matter, which he calls "the Primordial Nature of God," be guided in its evolutionary appetitions by "eternal objects" apprehended with a certain "envisagement." He expressly equates these "eternal objects" with Plato's World of Ideas. (*Process and Reality*, MacMillan, NY, 1968 (1929), pp 58 & 60, and p.39).

[140] How we will recognize it as "divine," we are not told. We may presume it will be the elimination of *death and suffering*. But since Christians claim that through the cross, suffering and death have been made the *tools of holiness and incorporation into the Logos* — i.e., "divinity" — why should these marvelous instruments, allowing humankind to participate in the redemptive work of the *Logos* itself, ever be dispensed with? According to Christianity's own criteria, therefore, why isn't the world, exactly as we now know it with all its suffering, the ultimate manifestation of the divine presence? *Existence*, by these criteria, may already be as "divine" as it ever needs to get.

where that we have seen it through the 14 billion years our universe has been evolving. In this case there couldn't be a fixed term to the *process,* or else it would mean either *existence* somehow achieved permanence and was no longer driven by the **conatus** (and thus it would stop being itself), or an outside guiding influence (which can only come from an essentialist source i.e., something *other than material energy* and its process) entered and changed the conditions of *existence* radically. For a fixed term to be true, *existence* would have to be transformed into something other than *the groping* **drive to survive** which is what we know and experience in this world everyday.

The implication of our analysis is that whatever end term there might be, the "Omega," like all integrated functions, *must itself be only a stage in a process.* It cannot really be an end point at all. It must simply be a plateau from which the process continues building ever "higher" and creatively unplanned developments. We can validly project such a future because we have a past of 14 billion years to go on. *Matter's energy* has constructed integrated functions, not once or twice, but over and over and over again, evolving untold numbers of complex organisms of astounding range of action and depth of ability. Each of the plateaus achieved by this process is an "advance" because, while incorporating the past achievements of the expanding totality, including that of the earliest and most remote ancestors, they have always gone beyond all of them to produce something unforeseen. But. nevertheless, in every case, *it was only a plateau,* not an end — permanent *existence* has never been achieved. And to claim it will be is a pure projection.

To the contrary, from the mountain of invariable past observations we can reasonably predict: *there is no end. There can be no end, or existence is not the process that we have observed it to be.* The process will end only if existence stops being itself — i.e., *material energy* wanting to exist, and creating time by doing so.

Such an emergent "Omega" would never transcend the material, and would be thoroughly *a posteriori.* But it would *always* only be the current ultimate product, the *integrated function* of all the structures and mechanisms that went before it, drawn into *temporal existence* by the self-embrace of *matter's energy.* Like everything else in our world this temporary "*pleroma*" would really be produced by all the agents of evolution. And even if this pleroma is "divine," it is a *material* divinity that never ceases creating *time* as the by-product of its labors.

If the goal of existence is to exist, please note: it achieved its goal from the very beginning *and at all points along the line.* For it follows that such an evolution, ironically, *is no more present at the later stages than it was at the beginning, or at any point along the time-line of development.* So we realize to talk about a "divine term" becomes somewhat irrelevant. Any point along the way, no matter how primitive and undeveloped, is equally part of the "divine" process, *because the "process" is to exist* — and therefore every moment, every plateau, even our own, is equally existent and therefore equally "sacred," even as it remains equally focused on the future. And any term, no matter how complex, advanced, or "divine," is itself simply another step in the *process* — "better," more developed, perhaps, but no more "existent" or driven to survive. Thus should humankind be disabused of any temptation to disdain as "inferior" those life forms that emerged before us and share *being-here* with us at the present time ... or to ever think of ourselves as superior to others. We are all part of a single developing organism; if we did not have their shoulders to stand on, we would not *be-here* at all.

In chapter 3 we said time does not precede *existence,* it follows it. Time is the product of duration. Once we understand that time is the sign and footprint of the creative process the idea that somehow we should "get beyond time" and into "eternity" appears unattractive, and in fact absurd. The energy that makes things continue to be, radiates time as a by-product of its labors. To be in time, therefore, is to be to be "at the cutting edge," to be "where it's happening." Eckhart called it the eternal "Now." It is the moment of creation.

There is a profound consonance between our personal drive to survive and the ongoing creative work of the substrate *in which we live and move and have our being.* They are one and the same thing. Our *conatus sese conservandi* is simply the human reprise of the **self-embrace** of the quarks and gluons, the vibrating strings of *material energy* elaborating itself in time. Where else would anyone want to be except here and now in the flow of time, part of *matter's energy* as it assembles this stunning universe that spawned us?

These speculations have directed our attention to aspects of *existence* that we may have otherwise overlooked — aspects that later will have a bearing upon our definitive interpretation of the Sacred. We are all aware that the human person is an *integrated function* which has been traditionally interpreted as having already achieved a certain measure of "divinity." In our tradition, we have to admit honestly, this has been taken to be *not entirely* a poetic flight of fancy.

summaries,
conclusions and corollaries to part 1

1. *existence* is different from "being." The traditional concept of being is associated with the dualism of matter and spirit as two separate and irreducible *genera* of reality. *Existence,* on the other hand, is one thing: a dynamism for maintaining itself ... an enduring material process. It is a dynamic self-embrace, not an *idea* bearing a purpose beyond itself. It is not an energy for *doing something,* but for existing. It is an energy for remaining what it is ... which, from that point of view, one could say, is an energy for *doing nothing and going nowhere.*

2. metaphysical individuality is a derivative of essentialism. Without "essences" there is no metaphysically grounded individuality at the level of the organism because there is no "principle" of existence besides matter. Matter is a shared commonality. Philosophically speaking, all we can say for certain is that a functional individuality exists. That means that we cannot eliminate the possibility that there is only "one thing" in existence: this universe with an *existential* dynamism that unfolds into everything that is. Participation is genetic, not idea-based, and supports the unity of all things.

3. *material energy* has two characteristics: (1) it is a living *existential* dynamism, with (2) a universal communitarian strategy.

4. *matter's energy* is "one thing." The heretofore separate notions, *material energy, spirit and existence,* are in fact simply different names for the same "one thing." It means that the nature of reality is *not* grounded in some other "spiritual" world where the conditions of existence as we know it — finiteness, temporality, extension, communitarian re-combination and change — do not apply and remain unexplained. For us, *existence* becomes directly intelligible as the very same material reality exactly as we find it in our material world and perceived by our material senses. There is only one "kind" of being ... and it is the *existential energy of matter.*

    This identification of reality as "one thing" means the traditional terms, matter and spirit, no longer apply. Our perspective says that at the heart of all things there resides *an intrinsic dynamism* responsible for all subsequent development including life, animal consciousness and human intelligence. Against the reductionists we have said matter is not inert and intrinsically passive and static ... and therefore change is not random and unintelligible;[141] *material energy* directs itself toward an inchoate *end*, primitive though it may be — *continued existence*. This dynamism with its foundational existential teleology means that the randomness we observe, even though universal, is at a secondary level. It doesn't explain it all, there is a remainder: a primordial thrust set in motion by the *congenital self-embrace of existence* collectivizing itself in order to survive. The only tools it has to work with is its own gathered self. This was in evidence within the first second of the big bang. Following Spinoza we have called this self-embrace **conatus**. It is an intense **"desire"** that does *not* seem to be a matter of evolutionary selection, but rather is presupposed as the dynamism that explains it. All things, no matter how primitive they might be, have it and have it intensely. In pre-living forms, it is displayed as a tendency to aggregate, integrate, achieve a stable identity and perdure. Among living things, it is the lust for life, the drive to survive that responds to *survival information*. In all cases it is fundamentally collective and cumulative: it forms bound relationships of earlier bound relationships. It can reasonably be attributed to the substrate itself. It is the very nature of *matter's energy. Being-here is a collective dynamism for being-here. **The purpose of existence is to exist.***

> [**Author's note:** *"want"* or **"desire"** *is a descriptive metaphor. The living dynamism, however, is a datum. We use "want" only because we have no other word for it. We do not know what it is. All we can do is describe what it has become in us. "Want" is the way it displays itself in us.*][142]

This view implies there is no *outside* "designer," like the "Potter God," directing the evolutionary process and injecting an *outside purpose* into things, making them intelligible. Creationism would have to include the claim that conscious creative design has been engineered in such as way as to **disguise itself as randomness.** That is patently absurd. *Existence* struggles because it is *on its own*. All change and becoming is ultimately intelligible in these terms.

---

[141] Cf Gilson, *Darwin ... op.cit.,* He says to explain the universe by "chance" is unintelligible.

[142] Arthur Schopenhauer, equally challenged to find a word to describe this phenomenon, called it "Will." For an extended discussion of Schopenhauer's analysis of the phenomenon and use of the term, and the differences I have with him, see the appendix.

From this simple empirical datum — which immediately becomes a self-explanatory "principle" — **all being and all becoming can be understood.** It is the ground and horizon of everything *we can know*.

5. the valid use of analogy. "Being" for the scholastics was composed of *essence and existence* and it was therefore believed to be distributed *proportionately* among things according to their essential level of "spirituality." Analogy was the epistemological reflection of an *ontological proportionality*. When the Thomists speak of the "analogy of being," they are principally referring, not to the epistemological reflection but rather to the alleged different "measure of being" in the things that exist.

According to what our analysis is revealing in this study, however, we say that what things *do* — how they behave — might be dissimilar, and therefore "analogy" may be appropriate for describing it, but we cannot say the same for their *existence*. Their *presence* to us, their *existence*, is *univocal*. **Existence is experienced as the same in every case — as presence — no matter what the level of function, whether pre-life, protist, plant, animal or human.** Since the Greek view was not based on scientific observation but rather on *logic* or "ideas," the attribute "existence" was gratuitously declared to partake of the "proportionality" of the "essence" to which it was subordinate, and so was considered to differ among its various applications *proportionate* (analogically) to the kinds of entities it modified.

There is proportionality in what *existence* **does** — how it behaves — depending on the level of emergence in which it is found. A way to conceptualize this *proportionality of function* is with *analogy*. But in this case, analogy is simply a description of behavior, a version of *metaphor,* not an existential "distribution."

6. *existence* and experience. While it has been traditionally assumed that there is *some* reality that is *not* accessible to experience (e.g., "spirit"), our analysis in this study rather supposes an absolute correlation between *being-here* and experience. Whatever **cannot** be *experienced* cannot be said by human beings to *be-here* (*exist*) and that for two reasons: first, because our perceptions are determined by and confined to matter (even the scholastics would agree with that), but also because, as I claim, *existence* is identical with *matter's energy* and therefore *all of it* is fundamentally accessible to the sensory apparatus (or its extensions) *designed* *by and for matter's energy.* (Human) consciousness is *material energy* looking at itself.

To the charge that we are establishing an exclusively materialist premise from which only materialist conclusions can be drawn, I would answer that every system of thought faces the same problem of data and evidence. They all have to begin with human experience which is bounded by the limits of materiality. To start anywhere else means assuming that "spirit" exists as a separate immaterial genus of being and that we are in touch with it. It must be recognized that the leap out of these initial universal conditions represents in all cases an inference or a projection. The validity of any conclusions about the existence of "*non-experienced reality*" must be evaluated by the validity of that leap. This starting point is universal. I would characterize my own position this way: I begin with the evidence of experience, but unlike the dualists or idealists I find no evidence that compels me to go beyond experience or its instrumental or mathematical extensions and recognize realities that "exceed the capacities of matter."

7. the survival of the "sacred." How is this new focus on *existence* (as opposed to "being") — this *integrative material process* — related to our inherited interpretation of the *sense of the sacred*?

The "**sense of the sacred**" in my view stands on its own as a human phenomenon — a common psychological and social experience. It does not immediately imply the "existence of God," as some claim. Nor does it appear to be derived from religious socialization; for people who do not believe in "God" also have a sense of the sacred.

Rather, in recognition of the intense emotional investment in whatever is considered sacred, it may be reasonably *understood* as a derivative of the **conatus**, our drive to survive as human beings. I will dare define it here: the *sense of the sacred* is a by-product of our existential self-embrace. It is the affective resonance of our appreciation of *our existence* — an appreciation which radiates out to everything that has to do with it, from its alleged source, to all those things believed necessary to maintain it.

[*Author's Note: Some readers may object to the use of the term "sacred" because of its religious associations through the millennia. I recognize that it is a problem word in this regard. I will gladly accept the use of another word or phrase for "sense of the sacred" so long as it continues to refer to a subjective feeling imputing an ultimate value requiring recognition. Its interpretation may be a matter of legitimate dispute, but the existence of the phenomenon is not.*

*Also I am intentionally bracketing the effect of society's collective appropriation of the **conatus'** energy to create **religion.**]*

So we *start with* a sense of the sacred as a human experience, and pursue an enquiry that tries to determine whether or not it has a justification that transcends cultural programming and personal predilection ... or to say it in a different way, an enquiry to determine exactly what may explain it, and what it, in turn, explains. Effectively we are examining the root and ground of the **conatus.**

Essentialist-spiritualist philosophy grounded the traditional "sacred" in two ways: (1) It said that "Being" was a spirit-"God" who designed and sustained our being with a *participation* in "his" being, and (2) that we human beings each had an eternal personal destiny with "God" precisely and only because we were "spirits" as "he" was.

The cosmo-ontology that we are proposing here affects each of those points differently. As far as (2) is concerned, eternal personal destiny was called into question because our position challenges the existence of the separable *immortal individual soul.* Personal destiny from now on will have to be calculated on a different basis, and with an entirely different result. To the degree that the *sense of the sacred* was tied to the (eternal) existence of the immortal individual soul, it is gone.

participation-in-being

But in the case of (1), participation in "God's" being, the question remains open. In the non-dualist view we are proposing in this study, the sacred is theoretically sustainable based on the "participation" created by the *common possession of the substrate, matter's energy.* What it comes down to is this: *material existence* as we have been studying it, performs the theoretical role once assigned to "God" as "Being" — it is that in which "we live and move and have our being."

How do these competing "grounds of the sacred" compare:

First, traditional participation suffers under an insuperable liability. It is premised, as we saw, on subsistent ideas. But there is no "World of Ideas" that makes traditional "participation in being" possible and there is no world of separable spirit. This will affect the "concept of Being" as the ground of participation.

The term "God" has been so wedded to the essentialist view that some feel it is impossible to use *the word "God"* without evoking essentialist spiritism. But the issue in this case is the word, not the reality There is no question that *material energy is **an existential factor*** of sufficient ontological power to sustain the *self-embrace* that gives rise to the **conatus** and our sense of the sacred. *Matter's energy* is indisputably that "in which we live and move and have our being" and therefore, *objectively,* can explain and justify the sense of the sacred.

Second, *process,* that aspect of *matter's energy* revealed and measured in *time,* is fundamental to our definition of *existence.* The basic "stuff" of reality is not a "thing," but a **dynamism** *with a non-rational intentionality, a self-embrace* for which rational consciousness is secondary, emergent, not antecedent, not directive. Anything built of it, therefore, will also be *a self-embracive process,* not an *idea* with a purpose embedded in a "thing." To the extent that "sacredness" was dependent on the presence of static essences wed to final causes (purposes) and possibly a "divine" terminus, an *Omega Point,* it is gone. What kind of "sacred" does non-rational *process,* reflected in the **conatus,** evoke? My answer: only itself, **an endless process of existing,** a self-embrace that is equally functional at every point along the timeline of development.

Third, we can say that a shared substrate that evolves all things suggests a participation that is material, genetic and thoroughly *a posteriori.* It is *not* built on an *a priori* plan moving toward an Omega; it's built on the aggregation of constituent parts, reproduction, *symbiosis,* a "genetic" *relationship* — family, community — the result of a *process* of invention and integration driven by an existential self-embrace.

If the energy at the base of matter — which I call *existence* — now performs the *existential* functions once assigned to "God," there is no reason, as far as I can see, why it cannot provide philosophical justification for our sense of the sacred.

"God" has always been considered "pure spirit." The energy of matter cannot be postulated of "God" without imputing *materiality* to "God." This is a critical issue for our tradition. That "God" might be material has been considered *almost entirely unthinkable* in the history of western philosophy. (But, see the appendix on the materiality of God.) The word "God" carries an ideological overload connoting "spirit."

Matter is a living dynamism ... does that make it sacred?

So let's bracket the *word* "God" for now. Hasn't **the function of the concept,** "God," in fact, been replaced with *matter's energy*? Aren't we constrained to call the source of our sense of the sacred, "the Sacred"?

The argumentation is this: the human sense of the sacred exists. What explains it? It is explained by a *conatus,* i.e., an irrepressible organic drive to survive that implies our love of our own *existence* and naturally calls everything that creates and supports it, unconditionally "good." But the *conatus* — the human drive for self-preservation — is no different from the life force as we find it existing everywhere in our world, in every species and in every substance, accumulated from the elements of the substrate itself. It is a homogeneous energy to which absolutely everything in the universe can be reduced. There is *nothing else!* Since we as humans, in our every fiber and function are nothing but this *material energy,* our sense of the sacred, which is our intense, irrepressible appreciation for our own *existence,* is justified and entirely explained as a derivative of *matter's energy.* Therefore it is *the substrate itself* with its *existential self-embrace* that can be called the source of our sense of the sacred.

But the *conatus* requires a recognition of its creative power that was in evidence at even its most primitive moment. Accepting the *conatus* as a **living dynamism** at the sub-atomic level, however, takes an *understanding* that transcends the information available to particle physicists working in isolation. Recognizing the homogeneity of the *dynamism* of the *conatus* across the levels of *existence* requires the use of a **retroactive interpretation** that looks at, not only what physics can directly observe and infer about the big bang revealed by particle colliders, but at what these particles are observed doing later on at virtually every level of emergence. The panoply of forms, pre-living and living, conscious, intelligent and purposeful, that result from the repeated application of the "stuff" and collective strategy initiated at the big bang, is exclusively built of quarks and electrons ... unless there is an outside "spirit," the *conatus* must come from there.

The evidence for it is clear. Its character as *existential self-embrace* is within us, and it is through the intimate "experience" of one's own *conatus* that it becomes more than a syllogism and overflows into a deeper *understanding* of all reality. But, that having been said, I want to emphasize, *it always remains a syllogism*:

M: "life" cannot be *reduced* to mechanical reflexes (i.e. there is a qualitative difference between life and non-life);

m: but our planet is teeming with life, and every living thing is constructed *only* of a physical substrate which on its own and in isolation appears absolutely lifeless.

C: *ergo, either* there is another, immaterial, source that introjects life into "matter," *or* the substrate, despite all appearances and reductionist claims, is itself a ***living dynamism***.

The syllogism is inductive and after examining premises and evidence concludes that "matter is a living dynamism" activated proportionately (analogically) as we have been saying. If it cannot validly do that, the argument fails, and the reductionist position holds, although always with a condition ... reductionism must itself, in turn, explain "life."

Please note: *I am not trying to prove the "existence of God" as traditionally conceived* ... the very idea of a separate "God-entity-person" disappeared with the disappearance of immaterial "spirit" and was only reluctantly acceded to even by mediaeval essentialists using "analogy" to justify calling "God" a "person" and not an impersonal force. I am rather trying to *understand* the meaning of the life-force, the source of my sense of the sacred. In other words, my question has changed. I am *not* asking "is there a 'God'"? ... or even "what is 'God' like"? ... but rather "what makes the universe sacred for me"? ... or, "what grounds, originates and explains my sense of the sacred"? This is an important difference, for if I slip and claim that I am actually discovering what "God" is really like (however true that may be), I have trapped myself by the "G" word and I'm back in the quest for *something that I claim does not exist,* viz., the Judaeo-Christian spirit-"God-entity," personal Designer-Creator, cosmic agent, punisher-rewarder and hovering provider of the OT "Book." The word "God" comes bundled with all these characteristics. This anthropomorphic "God-image," because of its long unchallenged history, resists metaphorization. And *metaphor* is the only valid use that imagery can be allowed to have. Once we use the word "God" we have a hard time conceiving alternative imagery.[143]

---

[143] It's important to emphasize that in this study I am trying to remain **strictly philosophical**. I am not rejecting religion ... how "religion" may respond to the new understandings we are discovering here is a separate topic altogether. By emphasizing the damaging power of the "G" word I am simply attempting to maintain the independence of a very fragile, easily derailed speculative imagination, which is the only instrument we have for exploring the sacred depths of reality as it has been revealed to us by science.

Once we stop looking for "God," as the cosmic agent imagined by our tradition and *understand* that "matter is a living dynamism" and accounts for every structure and function in the universe including our drive to survive and concurrent love of life, we can look at the sacred with altogether new eyes. It is quite different from almost anything that the mainline imagery of our tradition has considered to be "true" of "God."

It's the "God-lover" then who must decide how to deal with that.

8. creationism? Our conclusion that "matter is a living dynamism" may seem to approximate the position of those who believe they see an "Anthropic Principle" operating in the evolution of life in the universe. Some try to use it as a proof of **creationism.** Their argument is:

> ... since the laws of physics are perfect for the emergence of chemistry, and chemistry is perfect for the emergence of life, then it all must have been designed so as to yield life in general and human life in particular. Had any of the laws of physics been anything other than what they are, the universe would have been very different, and perhaps not possible at all, and life as we know it would not have evolved. [144]

To assert that such features were *imposed from without* by the work of a Master Mind and Craftsman is gratuitous; there is no evidence to support it. But we can (must) say what we see ... and what we see has produced a universe too vast to imagine with at least one planet teeming with a near infinite variety of life. Minimally it must be said, with Peirce, that we are looking at a living spontaneity, a ***living dynamism.***

Creationism is wed to a ***supernatural theist*** notion of "God." Practically speaking, that means a spirit-"God"-person who is a cosmic agent, who thinks and acts *rationally* (i.e., with *reasons,* for a purpose) on material reality from a spiritual realm beyond material reality. Creationists not only claim that the physical properties of the Universe were *specifically designed* for life by this rational "God," they also insist that direct divine intervention was necessary on multiple occasions *thereafter* for the emergence of phenomena like life, animal sentience, human consciousness and many other things. To my mind, this is absurd. The "anthropic" properties could not have been very well designed if subsequent interventions of a miraculous nature were still required to produce these emergent effects. On the other hand, to accept a "deist" evolution in which *existence* was *initiated by a rational Creator* and then abandoned to its own devices, would make the "anthropically designed" universe someone's little game,

[144] Goodenough, *op.cit.,* p.29

and the excruciating struggle for existence a senseless anguish needlessly extenuated over eons of geological time — all by the whim of an uninvolved absentee Parent. The projection is internally incoherent; for it is incompatible with the very notion of the omniscient, omnipotent, benevolent and providential "God" held by its protagonists.

The suggestion, on the other hand, that the primordial energy at the base and at the beginning of our universe may be described as an immanent, primitive, foundational, *non-rational intentionality* — **a paroxysmal self-embrace of existence** whose subsequent developments were all un-programmed self-elaborations, while not supporting the cherished image of a purposeful providential "loving Father," does admit the possibility of a *benevolent intentionality so immensely self-donating and non-particular* as to appear utterly "impersonal" to us. It can also correlate with the traditional characterization of "God" as **esse in se subsistens,** for *matter's energy*, as far as we can see, is neither created nor destroyed and appears to have no explanation beyond itself. This opening to pan-entheism (or pantheism) is sharply distinguished from traditional *supernatural creationist theism* on the following counts:

(1) **There is no *rational* consciousness** embedded in the primordial intentionality of *existence.* This is where we part company with Whitehead, for example, who claims that the "primordial nature of God" (which for him is the material substrate) is imbued with an appetitive "envisagement" of what he shamelessly equates with Plato's "world of ideas." But there is no other world, and no "mind" that constitutes it. The **conatus,** as we observe it across the levels of emergence, **approximates to desire, not to thought, purpose and plan.** And its "objective" is not a plethora of Platonic "essences" accounting for the "forms" of untold number of species, but rather one single common "goal" in all its emergent forms: **existence!** This non-rational, non-teleological character remains functional *without rational purpose* in every form *matter's energy* elaborates, no matter how primitive or developed. At the most primitive level there is, obviously, no *evidence* of any *rationality*; but even at the most advanced levels, as in humankind, it can and most often does **pre-empt and override a contrary rational preference.** For the **conatus** spontaneously rejects life-denying choices and suicidal intentions.

(2) **There is no plan, no purpose, no "point."** The only "purpose" is *to exist.* As a self-embrace *material energy* can't help *existing; it is neither*

created nor destroyed. **It has to exist.** I have already had the temerity to suggest on more than one occasion that in this vision *existence* displays itself as a dynamic material version of **esse in se subsistens**.

(3) **There is no creative action,** no "efficient causality" as from one entity to another (as, for example, from "God" to creation), *for there is only "one thing" relating to itself.* The physical-biological elaboration of the universe *is* (and doesn't just appear to be) **entirely immanent,** i.e., a self-initiated, self-sustained, self-contained and self-directed process. It is a self-elaboration, a self-extrusion, a self-unfolding not entirely unlike the way the oak tree rises from the acorned earth, or the way the rose unfurls its splendor.

Its **transcendence** consists in its ability to go beyond what exists at any given point in time and "extrude" new forms of existence from itself. Considering the "distance" covered from the first proton to the emergence of humankind, this transcendence is as beyond comprehension in depth and complexity as the physical universe is in size and volume. Infinite? Why quibble ... ?

is *existence* "necessary"?

With such an all-encompassing definition of *existence* as **esse in se subsistens**, haven't we come full circle on our initial *critique* of the concept of "being" in chapter 1, and now find ourselves ironically saying that *existence,* by being a self-embrace, is self-explanatory, self-subsistent *and therefore* **necessary** (and infinite)?

Our earlier *critique* of the concept of "being" was fundamentally a rejection of the ancient philosophical methodology which invalidly drew conclusions about reality *from an examination of concepts alone.* But, whatever we claim to know, cosmo-ontologyy insists, *must be directly observed and verified or be an immediate corollary to those observations. It is impossible to verify any necessity that is not a conceptual tautology ...* nor an infinity that is not a conceptual projection.

But please note: Cosmo-ontology is not thereby *denying* that *matter's energy* may be both infinite and necessary. Our rejection is as provisional as any other hypothesis. We cannot affirm it, but that doesn't prove that something infinite and necessary does not exist and that, perhaps, the totality of *material energy* necessarily exists ... and is infinite.

a living dynamism

If we were to classify "things" in an order of increasing complexity chronologically following the elaborations of evolution, we might come up with a "horizontal" chart that runs across the page from left to right in the following manner:

strings-quarks→protons→hydrogen atoms→heavy atoms→molecules→complex molecules→viruses→bacteria→eukaryotes→multicelled organisms ... etc, etc.

With such a schematic it is easy to think of these entities as distinct and separate from one another. One might then be tempted to imagine that life begins at a certain point on the chart, perhaps with viruses or bacteria, the earlier entities obviously not being alive.

But this way of looking at things fails to illustrate that the entities to the right *in every case* are constructed of *and include* those to their left. The more primitive are *structurally integral* to the more complex. A vertical chart would display these cumulative inclusions more graphically to show clearly that all things are simply extensions of what went before and ultimately only varied combinations of the particle-energy substrate at the very base of the pyramid. [145]

This is why *reductionism always remains an option*. Every part of every thing is made *only* of quarks and electrons. The *very same quarks*, with the very same "spins," "colors" and electrical charges exist in the protons of hydrogen and oxygen atoms whether they're found in the fusion furnaces in the heart of stars, or in a molecule of water in a muscle cell in my heart. The "quark in my heart" is neither more nor less than a quark; but that quark is **me**! These quarks of mine throb with life ... where does that life come from? *Either there is **another source** of life, like a separable soul providing life to my quarks from "outside,"* or the life comes from "inside" the quarks themselves which have somehow cobbled together a set of interrelationships so clever and powerful that they can activate potential life and thought and love! For, by our science, ***there is nothing there but quarks.***

From our observations, then, all life forms including ourselves are constructed out of untold numbers of living cells, that are themselves formed from aggregates of complex molecules, and those molecules are combinations of the many atoms built up from the simplest one proton hydrogen.

---

[145] see appendix on Molecular Evolution for such a chart

Entering the proton opens us to a *nano* world of particles, too small to see or manipulate, where the foundational stuff of atoms — quarks and electrons — are a form of the primordial energy responsible for everything that exists in the universe, whether inert or living, infinitely large or infinitesimally small — everything*!* The manifestations of life with its fierce desire to *be-here* that we are familiar with on earth have apparently drawn their energy from this energy substrate of the universe. As life complexifies and intensifies through the levels of evolutionary development, one thing seems to remain constant, ***an existential self-embrace***: a raw, implacable, insuppressible *existential dynamism* — the drive to survive. Unless someone would *unscientifically* attempt to insert an arbitrary wall of division between living things and the substrate out of which they are constructed, we have to say that life reveals that *matter's energy* itself is a living dynamism in which "we live and move and have our being."

9. summary

We might say that since the significance of *being-here* (*existence*) is established *in all cases* exclusively from its apprehension in experience, it is qualified by the constitutive role of the conscious organism (the human being), which was evolved by and for the self-embrace of *matter's energy*. From such an *endo-existential* etiology, we should expect little more than existential tautologies. Human consciousness is *material energy* looking at itself. *Existence* is nothing other than our experience of *matter's energy*.

In the ancient traditional usage, on the other hand, the ersatz significance given to "being" was believed to be established not from observation but rather from its conceptual characteristics *derived from another world* and were considered more real than material existence itself. The exchange of the one perspective for the other reflects the philosophical shift from the ancient / mediaeval vision of rational divine spirit, creating fixed permanent immaterial essences, based on eternal ideas, terminating in a fixed, eternal divine unity as finality, ... to the world-view suggested here, of *material energy*, in a process of blind, purposeless *existential* self-embrace, utilizing integrative recombination (community) as a tool of creative development, anticipating an unprogrammed process without term. If the keynotes of the earlier view of the world were *immortal living spirit, eternal idea, fixed essence, pre-determined static end,* those of the vision proposed here are *undefined existential energy, groping self-embrace, temporary phenomena, endless unprogrammed process.*

The word and concept "being," developed within the essentialist world-view, performed the functions for which it was designed. The view of the world revealed by modern science and cosmo-ontology, on the other hand, requires a different terminology and concept to refer to "being." We have chosen *existence, presence, being-here,* which we equate with *matter's energy.*

Since our analysis of "being" in Part One involved a critique of the classic philosophical use of *concepts,* it seems appropriate that we should proceed to examine the way we generate concepts, traditionally called *abstraction.* After all, the ancient interpretation of the significance of abstraction was one of the key factors that produced the dualist World-View of inert matter and vital form that shaped Western thought and civilization.

Part Two, therefore, deals with epistemology.

part 2
knowledge and understanding

preface to part 2

Epistemology is the study of knowledge. Knowledge has been a critical part of the foundation of the dualism of Western thought because it was believed that *knowledge* displayed characteristics that could only be explained by assuming the existence of immaterial "spirit."

Concept formation, in the dualist view, was also thought to put humans in immediate contact with the "Mind of God," for by knowing "essences" humans claimed to know exactly *what ideas* "God" inserted into things and what their purpose was. And it was the assumption that the human mind could clearly discern the *purposes* of a rational "God" that convinced the Church it knew exactly what "God" required of us. It became the basis of the elaboration of the "natural law" to supplement what was commanded in the "Book." Thus epistemology, metaphysics, law and religion historically have been intimately linked. The immaterial essences were matched by our ability to grasp them with an immaterial mind.

But metaphysics has changed. Cosmo-ontology sees the world in different terms. There is nothing real but *material energy* and no "purpose" except its survival. We can expect that our new interpretation of how human understanding comes to grasp the fundamental structures of *reality* will, similarly, be understood in terms of a profound correspondence between the two. For in *understanding*, material energy is looking at itself.

This second part of *The Mystery of Matter* will explore the deep connections between human consciousness and the *material energy* in which it congealed, and that constitutes it. As Aristotle taught us, there has to be a sameness between knower and known for knowledge to occur ... and the sameness, I claim, is had in the material substrate, the common possession by all things of *material energy*.

chapter 5
conceptualization and abstraction

human consciousness, human concepts

L et's take a fresh look at what's been called *abstraction* and how it constructs concepts. I will begin by presenting a plausible but hypothetical version of the structure of human knowing that contrasts sharply with the traditional dualist-essentialist theory. Later on in the chapter I will present evidence that builds toward the verification of that hypothesis.

I contend that the process by which we form concepts is a supremely *practical* mental operation whose function and purpose is to help us defend ourselves and *survive*. It is entirely organic. It is an evolutionary adaptation that in its fundamentals we humans share with other animals. Concept-formation was not originally and therefore it is not properly speaking a "scientific" or "contemplative" activity. In other words, it is not, as the ancients believed, a "seeing." It is a survival operation. It is not directed at *what* things are apart from their bearing toward my *existence*, my *survival*.

Also, human knowing is not "abstract," in the traditional sense, for it *does not transcend space and time*; nor is it instantaneous and infallible. I propose that *conceptualization is a process that aggregates individual concrete sense perceptions to produce universal or general images.* Human consciousness represents a more efficient ability to identify, classify and thus *consolidate* similar perceptions of antecedent and consequent events in the flow of real time which is the keynote of all animal consciousness, and vital to *survival*.[146,147] This consolidation is simply an in-

---

[146] Cf Edward O. Wilson, *On Human Nature*, Harvard U.Press, 1978 p.2.

[147] The interpretation of abstraction presented in this Chapter has been anticipated by both Schopenhauer and Bergson. Copleston indicates the similarity of Bergson's ideas to Schopenhauer's without suggesting a direct influence. I claim the same thing for myself. I discovered these parallels with the earlier philosophies in subsequent reading. I believe this independent concurrence

stance of the perception of the "one and the many" that we encountered in chapter 2. It is a tool of survival — for *being-here and staying here.*

The process by which consciousness generates concepts basically has three phases. The first two, selection and classification, apprehend the object of experience, and the third, language, stores and allows it to elaborate what it has learned.

First, consciousness *selects* or "snap-shots" that feature of an experience that is most salient for survival. It is therefore, in the first instance, not a generalization but rather a *selective reduction to singularity.* The subject perceives a lion but is not impressed with its tawny color or regal mane, much less where it is located in the Linnaean classification of family, class, genus and species. The perception focuses on the lion as a survival challenge. "This thing wants to eat me"*!*

The second, generic aspect of concept formation, the feature that gave rise to its appellation, *universal,* is subsequent, and represents *learning, i.e.,* the *recognition of similarities* among multiple perceptions. "This one is like the other lion that ate grandpa." The construction of universal concepts is the aggregation of like perceptions; it is the ability to classify and later recall identified similarities among the selected features of discrete experiences in time. Animals also universalize their experiences and form concepts. The lion *knows* I'm a potential meal; he remembers earlier ones; he salivates in response. It's a trait common across the phylum.

Memory, Aristotle astutely observed, is intrinsic to this process because concepts are generated from repeated experiences.[148] They are fundamentally time-related and concrete. *I deny that concepts are exclusive to humankind;* and I deny they are the instantaneous insights of universal eternal "essences." Concepts aggregate like concrete experiences and represent the gathered plurality with a *concrete image.*[149] It is essentially a time-based *process* of collation so that practical judgments are made based on the similarities of multiple experiences. Without it the an-

---

speaks to the obvious nature of the common content. cf. Frederick Copleston, *A History of Philosophy,* Vol VII, "Schopenhauer (1)," Image Books, p.34ff., 36.

[148] In the very first paragraphs of the *Metaphysics* Aristotle says, *"Now from **memory,** experience is produced in men; for **the several memories of the same thing** produce finally the capacity for a single experience."* Book A, 980[b], 25, in McKeon, ed *Introduction to Aristotle,* 2nd ed., Chicago, 1973, tr W.D.Ross, p.277 (emphasis mine).

[149] Cf Ludwig Wittgenstein *Tractatus Logico-Philosophicus* 2.1ff. Ideas are pictures.

imal organism would have no way of *remembering* the connection that exists between an event, "the lion appeared," and its consequences, "grandpa got eaten." It could never identify a second predator as generically the same as the first. It would never survive.

words and concepts

The third phase is exclusive to humans. It involves language. Language is the use of arbitrary symbols (generally words) to stand for concepts. It makes it possible for consciousness to use pre-stored classifications, *word-labeled* concepts, to serve as the source of future selections among later experiences. It also makes possible the subsequent perception of the connections among them. *Animals do not have words.* In humans, new perceptions are assimilated to older word-labeled concepts thus making recognition faster and more efficient for survival. It's a shortcut on the learning process.

Language becomes the property of the community. Words are remembered and passed on to offspring as pre-sorted information that the child utilizes as her own. Animals are not capable of this level of storage, recall, accumulation of information (learning) and anticipatory (preventive or productive) action.

Human conceptualization, which is already selective for survival in its raw form, becomes even more so when ossified by stored linguistic classifications that are part of what we call "culture." *Language,* tends to "totalize" itself, forming a universal horizon, pre-empting new perceptions or new angles of vision with selections that have already been classified and definitively *labeled* by the community. Language, from being an ancillary tool utilized by people to help manage and share newly perceived concepts among themselves, increasingly assumes the authority of tradition and tends to anticipate and pre-determine perception. Over time and through familiarity, people forget that words are arbitrary symbols; they come to be taken for the reality itself. And whatever does not have a word-label, tends to be taken as non-existent.

This pre-determination is of great practical importance. In the case of the concept "lion," for example, early recognition for the young human is crucial, because this particular animal is programmed to kill other animals like us and eat them*!* The ability to recall and recognize a label like "lion" is essential to survival. The community is invested in providing the necessary information about such predators to its youth well in advance and *in* *the complete absence of personal experience.* The prior experiences of others are represented with a symbol, "lion," and the appropriate responses can be put into action without the individual having to process the concept from scratch (as it were). The transmission and appropriation of these pre-processed classifications insures a high degree of survivability precisely because of the conceptual homogeneity (common language) held by the members of the local community affecting perceptions, judgments and responses to a wide variety of experiences. Through language, in other words, *the human community becomes essential to individual survival.* This is fundamental. Society built on language recapitulates at the human level the **communitarian strategy** that characterizes all survivability (= *existence*) in the universe.

plato's traditional world

Our hypothesis stands in stark contrast with the *traditional* claims made by the theory of conceptual universalization inherited from Plato, Aristotle and Aquinas, which has been the accepted wisdom in the West for millennia. Our proposed model implies a complete rejection of the classic view.

The traditional hypothesis asserts that conceptualizing consciousness is *exclusively human*; animals, they say, do not form concepts. Classic epistemology claims human consciousness *reaches through and beyond the singularity apprehended in sense perception* to elaborate the *universal essence* in the concept. The traditional view, fundamentally constructed on the Platonic theory of *ideas* and the division of reality into matter and form, grants that the concept is derived from the singular sense perception, but insists that its particularity is immediately transcended by reaching the "immaterial" form. The concept, thus imagined, represents the entity with greater precision because it *instantaneously* "captures" its universal immaterial essence, its "quiddity," *what* it is — in Plato's world, the *idea* or essence, the basis of its "being," the *purpose of its existence.* Plato's *idea* is an instantaneous insight, it is *not* time-related; it does not represent a *process* of accumulation of data and learning. It is against this supposed *transcendent metaphysical horizon of the universal concept*, that the traditional theory then judges the perceived individual metaphysically, i.e., as a singular example of a universal essence: e.g., **this** is a **lion**.

In contrast, I claim the concept is the result of a *process* based in time and repetition.

My argument starts with the objection that the ancient theory only works within an obsolete cosmo-ontological universe constructed of "matter and form." In that system the concept only makes sense if there are such things as "essences" for the mind to "capture." The model concept for the Greeks came from organic substance, a living thing, like an animal or a plant, which was believed to have a clearly defined "essence" or "form" whose *purpose* was discernible. These forms were believed to be fixed and eternal, the unchanging reflections of an immutable "God;" therefore their apprehension by consciousness was instantaneous and resulted in a concept that was infallible and universal.[150]   Time and process were irrelevant if not inimical to Platonic idea-formation.

The Greeks extended the model of matter and form to apply to everything. This became the basis for explaining abstraction and conceptualization. The concept's "abstractness" resided in the fact that the "form" "transcended materiality," i.e., went beyond time and space and the sense-perception of individuality. It was the assumed inertness of "matter," responsible for this non-essential individuality, that has been used to justify claims for the existence of the immaterial. Traditional epistemology was born of and conforms to the outmoded scientific theory of matter and form.

We no longer believe that such a universe exists; there are no "forms." Even on its own terms, however, the model is defective. The concept of *form* might seem compatible with science if one were to approximate it to modern DNA as the "essence" of an animal or plant organism. In these cases, too, the "structural constraints" that work against mutations might be adduced as further evidence of a *fixed essence*. The theory might also be applied analogously to reproducible items of human manufacture, like chairs, tables, houses, etc., each of which is always made of the *same material* and fashioned with the *same design* and therefore might be said to have a fixed "form" or essence.

But the Greek model does not work for a host of other objects like rocks, for instance, or dirt or clouds, or color or emotions, sounds, qualities etc. *These "objects" do not have a specific "form" except in Plato's*

---

[150] Aristotle, and his mediaeval followers like Aquinas, believed the universal concept was infallible; error occurred exclusively in subsequent judgments which represented the attempt to re-apply it to concrete realities.

discarded World of Ideas, and therefore the concepts that refer to them cannot represent "essences." Whatever essence is claimed for them is the pure projection of *arbitrary human classification*, and motivated exclusively by human convenience.

For instance, if I call this object in my hand a "rock," I also recognize there is no *intrinsic* "rockness." I call it a "rock" because it is a convenient classification that embraces a host of characteristics among items of a similar kind. It "works" in practice. But rocks do not share the same *anything* except *the name*. Their origins, geological age, mineral composition, structural characteristics, are all different. Some are sediments that have compacted over time and pressure, some are cooled hardened lava, some are the calcitic remnants of ancient hard shelled sea animals. They have nothing in common except the arbitrary traditions of human nomenclature based simply on surface phenomena: appearance, density, size, etc., issues *decided exclusively by us and for our convenience.* They correspond to the categorical standards of a filing clerk. They do not represent the intrinsic "reality" or "essence" of the rock. A small round glob of ferrous metal is not called a "rock," and yet a piece of unrefined iron ore of the same size and weight is. Science might consider them both "essentially" the same thing — iron. But for us one is a "rock" and the other isn't. So there is no "essence" of "rock." The "essence" is the arbitrarily chosen human criterion of classification.

A similar analysis could be made of many other categories. And when we take a second look at our concepts of living entities or human artifacts, in the light of how we dealt with the rock, we realize that the procedures are really the same, for we re-classify them upon receiving new data. The *classification of selected features* is what is functioning here as it is in the generation of all "universals." Universals are *conventional labels* utilized for unifying a multiplicity of similar data, selected according to current and local evaluations of human needs. They are *conveniences*, initially temporary and tentative, conventionally accepted and arbitrarily assigned, stored in language, and tending toward a permanence over time supported solely by the ancestral authority of the language community.

conceptual re-classification

In the view proposed in this study, concepts are essentially *time-related*. They are the result of a *cumulative learning process* and they are entirely *a posteriori*. They gather and collate repeated singular experiences. This time-based feature means that concepts are always *open to revision* as new experiences enter the picture and modify earlier classifications. For example, the criteria used by paleontologists for deciding what earlier species of hominid were truly human has changed. "Tool use" was once a defining characteristic of humanity. Then, given the growing awareness of the wide use that animals of all kinds make of "organic tools," and especially the sophisticated tool use by primates, the standard for human behavior shifted to "the use of *modified* tools," i.e., tools that the user fabricated or fashioned in some way. But even this criterion is being superceded as primatologists are realizing that chimpanzees, for example, not only select but actually re-shape their tools according to the particular food-gathering problem they encounter and the available material.[151]

Biologists are continually re-assessing the species or the genus to which certain life forms belong. We don't think twice about re-classifying things based on new data about them. It is a mystery to me how this everyday phenomenon of re-classification could have co-existed with the unsupported essentialist claims underlying the traditional metaphysics of matter and form and its associated epistemology.

The conceptualization process is based on a convenience that serves our society's (current) needs. And I will claim that those needs — the needs of survival — explain the dynamic evolutionary origin of our intelligence and therefore its cosmo-ontological structure and the "metaphysics of knowledge."[152]

recognition

Some may object to the exclusively "materialist" description offered here for concept formation. After all, they say, the "recognition" that is characteristic of human knowing has been described, by astute observers

---

[151] Barbara King, Ph.D., lectures on Biological Anthropology, The Learning Co. 1999

[152] Cf, John Dewey *Essays in Experimental Logic*, NY, Dover, 1954, an unaltered reprint of the 1916 edition published by the U.of Chicago Press. In the "Introduction" p.30, Dewey proposes the central thesis of his "instrumental" theory of knowledge. "Thinking," he says simply, "is instrumental to a control of the environment." Even the very data that are used for the resolution of problems are not "raw." They have been "worked on" by science to elaborate a usable item of knowledge for the purpose of solving problems. Dewey's "instrumental" epistemology conflates with the theory of knowledge presented here.

like Wittgenstein, as "immaterial."[153] I take up "recognition" in chapter 10 where its role in *understanding* is examined. The "immateriality" I am talking about here is what our tradition has claimed characterizes conceptualization, precisely because of its *universality*. The ability to create *universals* is offered as a proof of "spirit" because "matter" is allegedly limited to singularity in time and space. It is this claim that I attack. The significance of "recognition" is important for our study, but human conceptualization does not solely account for it, *for the animals not only classify but also recognize*. Hence, if recognition is "immaterial," animals must also have spiritual souls.

The perennial insistence that there is an unbridgeable gap — a difference in kind and not only degree — between animal and human consciousness stems as much from an overestimation of what humans do as from an underestimation of what animals do. Ultimately it is the ancient dualism of our culture that prevents us from conceiving *"matter"* as a living dynamism capable of producing all these effects in an unbroken linear continuum across the levels of *existence*.

truth and accuracy

In the view proposed in this study, sense perception has a *selective function* because our conscious apparatus was shaped by evolutionary adaptation to respond to the demands of *survival*. The *survival drive,* our *conatus,* insures that those aspects of incoming data that bear upon present or future continued existence are actively selected. This selection operates within a community context. In our case, it is language-dependent. The word-categories stored in the community's collective memory take priority and tend to pre-determine what survival-orientated selections are perceived and chosen. Selection quickly moves beyond anything like pure disinterested contemplation of the environmental *gestalt*, and toward the practical decisions to be made within it. It reveals human knowing to have evolved as *a* communal instrument of life, not an individual tool of eternal truth. Evolution demanded that "truth" be judged by the survival success of the word-tools of the community, not by the individual's scientific precision or contemplative depth. If society's suggested selections were "accurate," the individual survived; if they weren't, she didn't. A con-

---

[153] Cf Ludwig Wittgenstein, *The Blue Book*, NY Harper Torchbooks, 1965 (1958), pp. 3,4,5. One may dispute the use of the term "immaterial" for the experience, or even W.'s intended meaning, but not the phenomenon itself.

templative intelligence that was transported by the grandeur of a brilliant sunset or a lion's majestic bearing while failing to heed the community's shrill warning that this animal is about to attack, can be considered fatally "inaccurate." There is little doubt that, in the context of life in the Pleistocene epoch where human consciousness emerged, natural selection would guarantee that those who were not attentive to the community on such practical interests would not live long enough to enjoy "reproductive success."

There have been an abundance of studies made on the human instinct to heed the warnings and suggestions of others. In our day, human suggestibility is the basis of commercial advertizing that, no matter how puerile they appear, are guaranteed to have an effect. Humans are "hardwired" to listen to others. Peer pressure is an aspect of the community orientation of the individual. Human survival is genetically communitarian.

Now it seems undeniable that the community's focus on *selected* aspects of sense perception, also skews them. For the "accuracy" spoken of here has little to do with the essence, or metaphysical significance, or complex interrelationships, or esthetic value of the perceived "objects of knowledge." The instinct to selective perception emerged under the implacable whip of collective survival and still functions within that order of priority. "Reality" conceived from any other point of view, we must realize, was irrelevant. A moment's reflection on our own experience reveals that the selectivity we exercise within the field of incoming impressions operates the same way. Our perceptions are always under the suspicion of being distorted by subjective and community needs ... *as they should be!* ... because knowledge is a tool of community survival. That's the "truth."

enter larger society

As the human species developed socially and provided more protection for itself, social conditions became the significant environment in which consciousness was increasingly called upon to function. The apparatus of knowledge, even though elaborated by evolution to operate in less congenial circumstances, in a protective human context came to focus on selections which could be made with more security, more leisure. It was at this point that consciousness could choose to re-direct itself in a more disinterested fashion toward the objects around it. And concurrently, the issues that bear upon how one survives in society itself, rather than in the forests and savannahs, would tend to increase in importance. Interests changed, but the mode of operation of the apparatus was ever the same. Human consciousness still actively *selected* according to perceived human need identified by the community's words, skewing perception. From a scientific or contemplative perspective, and from that alone, skewing is a problem. But from the perspective of the surviving organism, it's not a problem at all. Such "skewing" is essential if the proper selections are to be made; that's because consciousness is not for "seeing," it's for surviving.

So far the first, *selection* stage of the conceptualization process

In the following stages, according to our theory, universal concepts stored in memory for future use, simply compound the skewing for they become a fixed feature of human interaction. The concept is an internal expression that corresponds to an external "word" (which may be non-verbal but not "immaterial"). Language is that compendium of externally constructed symbols socially agreed upon to represent the aggregations of selected perceptions of known utility to the human community. It is a social system of representation, classification and storage. It signifies the conventionalization of earlier metaphors, as we will discuss in a later chapter.

essentialist universality and infinity

In the classic view inherited from the scholastics, however, the conceptualizing process is said to reach "the true metaphysical significance of the object known," namely its *essence.* Rahner says the "concept" receives its universality from *a faculty embedded in the human knowing process* which allegedly reveals the *a priori* infinitude of human intelligence and its fundamental orientation toward the "truth." The concept as *universal,* is claimed to be *infinitized* — given an absolute horizon by the human mind, against which the object known is fully *understood* for what it supposedly really is: a logically limitless possibility — a universal.[154] Thus it achieves the "essential truth," an accuracy unconditioned by practical utility. This is presented as evidence of the existence and character of the human "spirit."

---

[154] This reproduces the view of Rahnerian Thomism as presented in *Spirit in the World* (Herder and Herder, 1968 tr Dynch. It is the re-application of the ancient Aristotelian doctrine of the *nous poieticos,* the agent intellect, that Thomas says "shines the light of 'infinite being' onto everything it sees." The infinity in question is allegedly the human spirit. Cf p 132ff, p.222-3.

I object to the assignment of the qualities "absolute" and "infinite" to our universal concepts. It implies they are focused on "truth." Like everything else in our universe, concepts are the products of a *process,* which implies time and limitation. Our classifications do not reproduce "fixed essences" with their supposed eternal absoluteness. We treat concepts as temporary and tentative. Concepts correlate to our need, not to "reality." Human abstraction is an operation in which the individual experience is judged against a community classification (concept and word) shaped for the purposes of *survival.*

The provisional, temporary nature of our concepts is, for me, the clearest evidence that we have no right to claim that they correspond to "eternal essences" which has created the illusion that human consciousness is orientated by nature toward absolute truth. If we are willing radically to change our definitions of species when we receive new data about them, it is an indication that the concept was, all along, recognized to be provisional. The concept is a community label, the instrument of an ongoing process of living, not a metaphysical discernment.

the animals: evolution and the structure of consciousness

Animal consciousness is different from us in *degree,* not in *kind.* Close examination of the primates reveals characteristics we would be hard pressed to deny were, by the ancient traditional standards, "immaterial." Among them, tool use, complex communication involving linked strings of gestures, the ability to imagine the amount of information a companion might have (called "theory of mind"[155]), self-recognition, complex inter-personal relationships, and true concept-formation,[156] all point to abilities in the animals that have been systematically overlooked or underestimated. Classic epistemology has categorically ignored the significance of animal conaciousness and disregarded

---

[155] In her lectures published on video tape in 1999 by the Teaching Co., biological anthropologist Barbara King refers to experiments with primates that represented "dangerous" situations where one individual was able to see the danger approaching another who couldn't. Observers recorded that warning reactions were forthcoming in those situations where it was clear that the partner could not possibly be aware of the danger. In contrast, when it was clear that the second animal *could* see for herself, no warning was issued. She cites other observations from the wild where hunting partners have been seen anticipating one another's moves and intentions.

[156] Stephen Pinker, *The Blank Slate,* 2002, p.55 states that " ... [primates] are outfitted with many complex faculties which used to be considered uniquely human, including concepts, ..." and cites in support Gallistel (1992), Hauser (1996 and 2000), Trivers (1985).

any differences among them. Animals were all lumped together in one utterly reductionist classification — as non-human mechanisms, with all their conscious operations, no matter how complex, considered merely Pavlovian reflexes.

There is no dispute about our human ability to do more than any animal, intellectually speaking. But our superiority is *not* based on an innate watershed "thrust toward the infinite," as Karl Rahner proposes, but rather the *linguistic expansion* of the same ability to select and aggregate selections — to navigate the "one and the many" — which the animals also do at lower levels of efficiency and scope.

Animals do not have words, nor the social structures necessary to collect and house them. It is our verbal symbol-making capacity that allows us to generate and manipulate the words "infinite" or "transcendent" precisely as categories of classification. Words are *labels*. To call something "infinite" is not the same as conceiving or *knowing* the infinite. Even among traditionalists no one would claim that we actually *know* the infinite, except by inference (meaning by absence). The very word itself, "infinite," reveals its indirectness. To recognize the end-point of human knowing and to imagine a "beyond" which is *not* known but which we wish we knew, does not constitute *knowledge*. It constitutes a sophisticated *labeling* whereby we are able to "name" what we imagine. It exactly parallels our claims to "know" non-being. These are verbal tricks. It's always struck me as philosophical *legerdemain* to claim that to name what we do not know constitutes *knowledge*.[157] But, as we'll see later on, the very fact that we don't *know* may occasion a different cognitive embrace, one that may yield *understanding*.

We have the community's language as a storehouse of labels, so our greater memory capacity gives us access to what we've classified and can bring it forward for future use at will. All these differences, while admittedly *special*, are not *generic*. They do not amount to a "transcendence" over some gratuitously projected "limitations of matter," which is the claim of the traditionalists. They are *special* for they make us a *species*, but they do not make us a new *genus* of being.

---

[157] This is my objection to Rahner's approach, which reproduces the Kantian *a priori* perspectives.

other examples

While the primates are especially close to us, many other animals manifest an awareness of *aggregates* of perceptions which trigger an appropriate response. In other words, the animals' ability to recognize *an identity* even though accompanied by a considerable amount of variation, can only mean that something akin to our "recognition" is functioning for them as it is for us. My dog, for example, recognizes me with or without hooded winter coats, jump suits, full-face pesticide masks, seated on a tractor or in a car, walking, or wearing sun glasses, talking in a high or low pitch, singing etc., etc. And she is able to distinguish me from others so dressed. There is something operating in canine consciousness that goes well beyond a mechanistic reflex.

Animals form concepts like we do: by *learning*. I recently observed a young deer trying repeatedly to go under a woven wire fence. She was in the process of *learning* that the fence, by reaching to the ground, prevents entry here, there and, eventually, everywhere. But it was clear that she was not able to process these similarities immediately, for she kept trying in one location after another, all contiguous. She had not *universalized* her experiences and simply jumped the fence. A hunter would have had plenty of time to aim and fire at the fawn. "Intelligence" enhances survivability dramatically.

There is no need to have recourse to a transcendent "spirituality" to explain the *process of universalization.* We observe, in these examples, that animal consciousness is not simply a mechanistic reflex. Higher animals like deer truly *learn* by "universalizing," i.e., recognizing *over time* the similarities between discrete experiences[158] I believe that we are justified in saying, minimally, that their mental processes are *generically* the same as ours. That our abilities might be *specifically* different from theirs is not sufficient justification for claiming that we are of a different *genus of being* altogether.[159]

---

[158] Animals also over-generalize. Animals trained to expect a certain cause-effect sequence can be tricked by exploiting predictable behavior.

[159] Even the physiological processes that support animal consciousness are exactly the same as ours. Macaque monkeys, for example, have been proven to use the *same neural pathways* as humans in the cerebral operations that underlay visual representation. (Crick and Koch, "The Problem of Consciousness" *Scientific American Special Report* March 1997). And, as we all know, much research about humans is performed on *mice* and other animals because the location of organic structures and the relative involvement of all fundamental features, neural, hormonal anatomical etc., is the same.

evolution: the one and the many

We must not think of ourselves as evolving separately from the rest of the animal kingdom. We are a product, most recently, of mammalian evolution. Our ancestors in this line passed on to all their progeny certain foundational features of consciousness which we all subsequently enjoy. There are certain common characteristics of animal consciousness that we share and they derive from the "nature" of cosmic reality:

The first is *identity*. Animals, human and non-human, must "recognize" identity. A "thing" (including the subject of consciousness) perdures as itself and that phenomenon gives rise to the understanding of "one and the same."

The second is the *multiplicity* within that sameness: that there are (1) many things of the same type and that (2) the same identical thing exists in a temporal multiplicity of sequential apparitions, as we saw in chapter 3. This multiplicity within unity is the foundation of "the concept;" it represents a simple but crucial modulation of "one and the same" into the "one and the many" — the basis for "universalization" as well as logic and number. If there is a genetically encoded cerebral basis for "intelligence," it must necessarily exist in many species, because this ability to apprehend the "one and the many" is exercised pervasively across the phylum; it is not species-specific.[160]

This is not just of academic interest. It would be dangerous in the extreme for potential prey to be ignorant of the sameness of the many individuals within a predator species, even though the victim may have experienced an attack from only one of them. Species that survived *had to have* evolved a consciousness minimally capable of grasping the "one and the many;" they had to be capable of universalizing.

language and human intelligence

The specifically human version of this phenomenon makes use of the "word," — a "label," a symbol — to stand for a concept. Animals have concepts, but they do not have "words." Their concepts

---

[160] Stephen Pinker, *The Language Instinct*, Perennial Classics, 2000, p.244 asserts that "when an environment is stable, there is a selective pressure for learned abilities to become increasingly innate." The point is relevant to our discussion because it means that "hard wired" features of consciousness, like instinct, are not opposed to learning, it is rather the contrary: learning leads to "hard wiring."

remain internally image-bound for exactly that reason.[161] The communication that animals do — and it is considerable — is based on the simultaneous presence of the same internal image in more than one individual animal. There is no *discrete,* conventional external link between *animals.* Communication is accomplished by links that are the natural *organic* extensions of the internal image. They are the direct display of the emotions and reactions generated by the internal image. They are as smoke to fire; they are *natural signs.* Each animal recognizes the reaction in the other, and a stored common image is recalled which elicits a common response. Example: the presence of a mountain lion stimulates an externalized reaction in one deer (starting, snorting, running) that is immediately recognized by another, bringing forward in the second animal the stored image and stimulating the appropriate common flight. Also notice: these "signs" have an existential significance: collective survival. They necessarily put selective pressure for communication and the ability to "imagine" from symbols front and center. They help explain how species enjoying those traits have emerged and prevailed. They explain the human domination of the planet.

Humans also form generic images — concepts. But the labeling, the "word," that accompanies human conceptualization makes the concept capable of being organized and having influence on a new level — it complexifies human interaction and self-awareness exponentially. The symbol, the "word," unlike the natural, *concrete* organic signs of the animals, is a *discrete* (arbitrary) externalized sign that, precisely because of its artificiality, achieves a certain *reality of its own* outside of and potentially out of focus with the imagery it was originally chosen to represent. Words (and other symbols) are fabricated entities that ultimately constitute a parallel "world of words" just as "real" and available to selectivity and conceptualization as the world they are supposed to represent. The relevant insight is that there is no guarantee that language will accurately reflect the mental images it was originally intended to signify. It is here, with *words,* that the entire phenomenon of human intelligence — abstraction, imagination, error, and of course, deception — becomes possible and intelligible.[162]

---

[161] Ludwig Wittgenstein in his *Tractatus Logico Philosophicus (TLP)* famously suggests as a central insight that propositions do not convey some abstract "truth" through the links of logic but rather use words *as simple signs* to "show pictures" that are accepted by consensus and are thus said to "have" logic.

[162] Plato's "World of Ideas" is, in reality, this world of words.

First, by becoming a kind of "thing" in its own right, the "word" itself becomes a source of an *independent imagery* that may distort, skew or impose a slant upon reality. Language is the tool of human **imagination** *par excellence*. It provides a "distance" from present experience that allows the human mind to construct a reality that isn't there. The ability to imagine is perhaps the characteristic of human consciousness that most clearly separates us from the animals. It is responsible not only for our esthetic achievements, but for the transcendent accomplishments of humankind in virtually every field of endeavor: technology, politics, abstract thought, etc. It has allowed us to dominate the earth.

But the ability to *imagine* also allows us to *impose meanings and significance* upon reality that may or may not be there. These impositions can be subtle, and not necessarily negative. The word "to stand," for example, evokes an image of solid presence, and when connected with the prefix "under," serves to conjure up a picture with a "slant." To say "I understand" is a vertical image that suggests complete control, as if the observer were "standing under" assessing the "true weight" of the thing known. In a slightly different fashion, the word "com-prehend," derived from Latin roots, evokes a "grasp" that "surrounds" the object like a pliers also perhaps implying "control." In each case, the word provides *an image for an unimaginable object* — something utterly private and incommunicable. Some word is crucial, for without a word how would I "act out" or "point to" *understanding*? And even with the word, how can I be sure that *"understanding"* means the same for you as it does for me? Besides, do the evocations of "control" in each case really describe *understanding*? Or does it fool me into thinking there is more going on here than really is? Words give imagination tools to work with. Without them, imagination may still be able to function, but barely. The first thing wordless people do is construct a language — tools to work with.[163]

Secondly, in being assigned a "word," the individual internal image (concept) is converted into externalized *community property* and loses the concreteness it once had; it is no longer absolutely bound to the particular experiences it once gathered nor to the individual who created it. The word instantaneously achieves a measure of distance, an *abstractedness*,

---

[163] L.S. Vygotsky, *The History of the Development of the Higher Mental Functions*, NY Plenum, 1997.p. 173 ff.

which it never loses.[164] This distance of the "external" word from its associated internal concept makes self-awareness possible because it permits
human beings to "transcend" spontaneous identification with their affective
state and look at themselves looking, to see and hear themselves thinking, as it were "from the outside." It's a perception mediated by words.
Without words, human beings would be as locked into their spontaneous
feelings as the animals.

All words, because they are *discrete, arbitrary* symbols, necessarily
abstract from the concrete images that they represent. Words create the
original duality. They permit the existence of a parallel symbolic world, a
world to which — fortunately or unfortunately — we can relate on its own
terms as if it were an independent reality alongside the "real" one.

abstraction

As common property, then, the word becomes capable of evoking images in a multiplicity of hearers who may not have had exactly the same
experience. Thus it accomplishes communication even while it may skew
perception with frozen (and not necessarily accurate) communal imagery.
Abstraction — by which I mean the *distance* of a word from its concept-
image — accounts for the later, more "developed" ability to *imagine, to
fictionalize, to project,* and of course to misunderstand, distort, deceive
and lie. This is distinctly human. Animals are incapable of lying, because
they do not have *the discrete distance* on their concepts provided by
words. Abstraction separates us from our actual mental and emotional
processes and allows us to identify inclusions, subsets, exclusions and
hierarchical relationships of all kinds, including numbers, *based on words
alone.* Multi-tiered, simultaneous identities — *integrated functions* — also
fall within the purview of this capacity, derived from the foundational understanding of the "one and the many" now managed by the word-sym-
bols themselves creating the system of relationships we call language.
Such manipulation is the basis of logic as well as grammar.[165]

---

[164] The word "abstract" as can be seen from this usage, carries the radical signification "drawn
from" with a connotation of "distance." In this study it does *not* in any way imply, as it does for the
traditional theory, an immaterial essence.

[165] I believe the "computationally complex" grammatical rules claimed by Noam Chomsky to explain
the phenomenon to which he refers in *Language and Problems of Knowledge (op.cit)* are simply
examples of the hierarchical ordering of verbal symbols, as he admits on p.45: "*The rules operate
on expressions that are assigned a certain structure in terms of a hierarchy of phrases of various
types.*" Subordinate clauses are most often identified by relative pronouns which create an unmistakable inclusivity. They are functions of "the one and the many."

I don't mean to minimize the manifest differences between animal and human intelligence. I'm simply trying to show that the two phenomena spring from the same ground; the one is a much more elaborate — double-tiered — version of the other. The human ability to create and manipulate symbols (words and numbers) has made all the astonishing human accomplishments possible, including mathematical science and the technology that uses it. But it's important to recognize that *symbolization* (language) *is a function secondary to conceptualization* (the aggregation of the many perceptions into one consolidated image) and no matter how sophisticated it appears and astonishing its accomplishments, symbolization should be seen as **categorically subordinate** to the more generic operation of *universalization* from which it developed and to which it owes its character. I believe by these criteria, language with its *symbolically guided* hierarchical structures reveals itself to be merely a sophisticated sub-set of the inclusions that define animal intelligence — the capacity to understand the "one and the many," to conceptualize.

I would hazard a guess that if we were lucky enough to still have with us living examples of earlier sub-species of *homo,* now extinct for reasons that raise suspicion about the genocidal capabilities of our own sub-species,[166] we would have evidence of more primitive symbols standing for more rudimentary internal images. Such word / gesture structures would doubtlessly have revealed that the basic paradigms of language were elaborations of the same ability to understand "the one and the many" common to all animal consciousness. Simply because we are currently alone in our highly developed language capacity does not mean that language must always have functioned only according to the manner of our later learned responses. There had to have been more primitive versions of word usage in earlier times even with the same cranial capacity and apparatus.

Those who have been privileged to observe the linguistic development of very young children know not only how concrete, singular and groping initial language usage is, but how rapidly the usage complexifies and expands in ever widening inclusivities in an astounding manner. The infant brain is organically the same before and after these developments. It is not like some new organ or organic function suddenly kicks in. It is clear

---

[166] Cf chapter 6, pp. 170-171 and associated footnotes for a brief discussion of this suggestion.

that once given the appropriate word-tools, the youngster quickly masters complex logical, grammatical and numerical operations. This suggests that the same brain in a Neandertal or early human without access to the learning and reflexive hierarchialization that a fully developed language offers, would appear almost animal-like in its mental abilities in comparison.

So language is the key to intelligence. But if the hypothesis I am proposing is correct, no language, no matter how primitive ... or how sophisticated ... would function otherwise than in terms of the inclusions and exclusions of "the one and the many."

But this is not a mere fortuitous occurrence, a genetic accident isolated to the clade. I believe human intelligence, like all understanding throughout the animal world, *mirrors* the configurations found pervasively in the environment out of which consciousness emerged and for which its organic structures were shaped. The fundamental understanding of "the one and the many" characterizes animal consciousness *precisely because it characterizes all the extant elaborations of the substrate* in the animal's environment and the organism of the animal itself. Concept formation mirrors reality.

Matter's energy is on both sides of the consciousness divide. Knower and known are equally made of the same "stuff" and were developed by the same processes that have resulted in complex multicelled organisms that reproduce, creating evolutionary speciation. The "one and the many" recapitulates the fundamental communitarian structure of the substrate whose stages were surveyed in chapter 3.

reflexive self-consciousness

All human conceptualization is derived from and resembles "animal" recognition of similarity which is the basis for universalization. But having words means that humans can go further than the animals. We can "objectify" our thoughts with words, identify, examine and communicate *interior experiences* — the experience of experiencing, or the experience of conceptualizing, the experience of assigning and using words for concepts (metaphorization) — and thus achieve an overview and extended control over them. Verbally labeled mental operations are as open to observation, shared analysis as well as manipulation (e.g. *self*-deception) as is the observation itself. Reflexive oversight is potentially endless: we can look at ourselves looking at ourselves looking at ourselves, etc., etc. *ad infini-tum,* and communicate virtually all of it.

This *reflexive self consciousness,* made possible by the use of words, has often been adduced as evidence of a distinct, unique and transcendent capacity which was claimed to give us a direct contact with our immortal spiritual souls, the alleged source of our transcendent "personhood." We had not only *consciousness, but self-consciousness.* *Self-consciousness,* the awareness of the "ego," was believed to make humankind not just a different species of animal, but a different genus of being altogether: *immaterial spirit.*

Experiments with chimpanzees, however, have shown that by the simple act of preening themselves in the mirror, these genetic cousins of ours indicate that they, too, recognize themselves as "selves," and are not confused that they may be seeing another chimp in the mirror — as birds do when they fly into and peck at windows.[167] This suggests that for chimpanzees, even without words, *the interior observation of their own mental processes must be available to them in some form.* Until they tell us, of course, we'll never know for sure. But their behavior indicates that what they do with their minds is very similar to what we do with ours.

conclusions: conscious operations beyond "survival"

On the basis of the traditional misinterpretation about human *versus* animal consciousness, we have become accustomed to speak of human intelligence in transcendent, that is, *immaterial* terms. The interpretation offered in this reflection, however, claims that the operations of consciousness at the level in which they are considered most characteristically human — abstraction and universalization — are simply the verbal expansions of the image and memory capacities possessed by all animal consciousness. Conceptualization functions, for the purposes of survival, to classify selected aspects of the raw unorganized flow of realty presented to experience. That means it is not naturally "disinterested" and "truth" is not its object. It is clear that, initially, concepts bear little reference to any "scientific" or philosophical interest about *reality-as-it-is.* They do not focus spontaneously on speculative questions, nor on a precision

---

[167] Edward O. Wilson, in his book *On Human Nature,* Harvard U.Press, 1978, pp 26-27, commented on such experiments with chimps and mirrors carried out by Gordon G. Gallup. Cf a NOVA program on Public Television in the late '90's.

that has no apparent value. But, thanks to language, we have the ability to observe our observations. The conceptualization function can be *monitored and intentionally re-directed* to work toward a more "objective" representation of reality — one that is not dominated by the urgent demands of survival. Our abstractive abilities may have originated from survival and self-interest, but they do not have to remain that way. The salient point, however, is that the scientific or contemplative use of consciousness is *an* **exaptative**[168] *application* discovered and chosen by us; it is not "natural."

I want especially to mention the simple contemplative appreciation of our co-existence with the things around us — *being-here* in its concrete manifestations. This capability of embracing reality in an act of disinterested consideration is regarded as a most cherished mental function, and was once thought to be an essential quality of our intelligence, another item that supposedly confirmed our "spiritual" nature. This contemplative function, however, like our highly refined scientific procedures, is a *learned and chosen* operation, made possible by the leisure provided by the human community. Our contemplative faculty will come up again in the next chapters on relatedness within *being-here* and the metaphorical tools used to apprehend and express it.

This is central to our perspective. The more refined and disciplined uses of human intelligence — those that precisely go beyond the spontaneous instincts for survival — are a human development, not a divine creation.

Human consciousness is *not sui generis.* It does not transcend the *genus* of animal consciousness. It is, on the contrary, determined by the ancestors from whom it emerged and whose character it conserves. As an organic function, consciousness does not derive from any transcendental reality that stands above and apart from the process and occasion of its emergence. It does not, in other words, bear any necessary reference to the consciousness of a Superior Spiritual Being, or to any abstract quality, like immaterial "Mind" for the existence of which the "immaterial" operations of human consciousness have traditionally been offered as evidence.

Human survival in the future may occasion the evolution of an even more expanded power of abstraction and classification, whose operations

---

[168] a neo-logism of Stephen Jay Gould that describes a trait that was evolved for one reason and gets used for another.

at this point we may not be able to imagine — just as the deer cannot imagine what it means for us humans to think as we do even though our intelligence is simply a linguistic expansion of theirs. But we can categorically affirm, that however astonishing these hypothetical future abilities may appear, they will similarly *not transcend materiality* nor our animal origins; they will always be based on "the one and the many;" they will always generalize singular perceptions. Our consciousness as it now exists has been organically structured by its past in a universe of *matter's energy* — the animal forms and the environment from which it emerged and to which it adapted — not by the alleged heavenly world of spirit where it was claimed we are going. Similarly, whatever future state emerges from the evolutionary process will be determined from what we are now and what the environment and social conditions — increasingly created by us — will require.

chapter 6
mind-body: the modern dualism
the somatic dimensions of consciousness

1. consciousness and the body

Our traditional theory of knowledge was a derivation of the Greek separation of spirit from matter. "Knowing" was considered an exclusive function of the "mind" as opposed to the "body." And the mind was identified as immaterial spirit. René Descartes in the mid 17th century accepted Greek dualism but gave it a twist inspired by the newly emerging experimental sciences. He saw matter as an independent substance, a "thing" he called *res extensa* with its own properties independent of any "essences." He claimed that the passive inertness of matter, described by Chomsky as "contact mechanics," clearly *proved* that there had to be a "second substance" separate from matter, that actively *did* things, whereas matter could not. He called that second substance *res cogitans,* "mind." Today, even in a reductionist context, the accepted wisdom is not much different. The view that matter is dead and lifeless means that "knowledge" is still considered a profound mystery, somehow incompatible with the "matter" of the *human organism*, and gives rise to a "mind-body problem:" i.e., how can these two irreducible "things" co-exist and interact. The illusion of a mind-body duality continues to dominate the interpretation of being human.

The position taken in this study, however, is that conscious activities are organic events that occur *within* an organic environment. They are all explained by the living dynamism that is organicity itself. There are not two separate irreducible "things," there is only one — *matter's energy*. Cognition is a valence within *matter's energy*. *Matter's energy* is the common ground of this experience, its content and driving force. By granting the exclusivity to *matter's energy* of which the human body-person is a modality, we establish human knowing as a subordinate, relational event within the material substrate. Whatever had been called "spirit" we now know to be *a function of matter's energy*. So, in our view, there is no possible duality; what were called in the past, "matter and spirit," are one and the same thing — *material existence* experienced and measurable, residing on both sides of the conscious relationship of knower and known. This suggests the "mind-body problem" was created by our multimillennial imaginary western dualism, and will disappear with it when it goes.

I reject the "immaterial" conceptual origins traditionally offered for the interpretation of human consciousness and its apprehension of "being." Once the false primacy assigned to "immaterial" conceptualization has been thoroughly exorcised, we will be well on the road to correcting this fallacy which is more than an academic epistemological error. For the illusion of *the disembodied human intellect* is the origin of the distortions of western cultural mind-set: an arrogant and self-destructive individualism, the denigration of the body, an overweening rationality, the license granted to runaway technology and, historically, the justification for the exploitative colonization of people around the globe less obsessed with abstract cerebration and its power projections than we.

Evidence of the nature of our immersion-relationship to *matter's energy* is had in the suffusion of the operations of consciousness throughout the entire organism, the body. There is no "mental" activity that is exclusive to the brain or the cerebral complex. Thinking is an inclusive activity in which the entire organism is engaged, even if all parts of the body not to the same degree.

> The 'mind,' as the term is used more technically in this article and in the philosophy of mind in general today, encompasses a variety of elements including sensation and sense perception, feeling and emotion, dreams, traits of character and personality, the unconscious, and the volitional aspects of human life, as well as the more narrowly intellectual phenomena, such as thought, memory, and belief.[169]

There is also no evidence for any human activity, intellectual or otherwise, that does not *reside entirely within the human organism, the body, which includes the brain.* Thomas Aquinas' famous *dicta* that "all human understanding is exclusively focused on the sense image,"[170] represents an acknowledgement of this same fact. It is a fundamental canon of our view of the world.

---

[169] From, "**mind, philosophy of,**" *Encyclopedia Britannica,* from Encyclopedia Britannica 2006 Ultimate Reference Suite DVD
[170] ST I, 84, 7.

Consciousness is a valence. At all levels of life it is the instrument of the connectedness of the material organism to its material environment. Human consciousness is no different. It is primarily characterized by its organicity which is the basis of its relations with its surroundings as a function of the **conatus,** the survival drive. Hence, the *somatic* dimension, the role of the body, often ignored or at least minimized in considering the characteristics of human consciousness, is its foundational and predominant feature. Abstractive verbalized conceptualization is a highly specialized and relatively unused ability that functions as a surface feature riding on a sea of non-rational, or pre-rational non-verbalized corporeal relationality. And it is that *somatic relationality* that primarily characterizes our cognitive connection with the things around us.

Unfortunately, the exaggerated enthusiasm of the Greek philosophers and their Cartesian disciples for ratiocination and its logico-technical potentialities tended to be given center-stage in the evaluation of human consciousness. The somatic ground was at least de-emphasized, but more often it was openly disparaged as being on the wrong side of the dualist divide: a function of body, not mind. The discoveries of this study have changed all this. *The somatic dimension is fundamental* and is always operative suffusing rationality with its non-rational perceptiveness. It is the engine and defining category of human consciousness. And, as I will try to elaborate in this and the following chapters, it is the basis of an *understanding* to which "knowledge," defined as verbalized conceptualization, is subservient and secondary.

2. the awareness of the *ego*

The first fall-out of this emphasis on the somatic dimension in knowing will be an new appreciation of the character and role of "self-consciousness" — the "transcendence of the ego" — in human experience.

Experience is an interaction within the totality of *matter's energy*. It only secondarily yields awareness of the "I." There is an organic homogeneity not only within the knower between mind and body, but also between knower and known. There is no *reaching* going on here. There is no *bridging*.[171] What there is, is an expression of the non-mediated, direct self-embrace of *existence* as the all-inclusive envelope, the context in which consciousness occurs. Ego-consciousness is an event emerging within pre-existent sameness, *matter's energy* — the shared organic reality of the interacting entities, knower and known. *There is no "metaphysical" difference between them because they are all modalities of the same substrate*; therefore there is no mystery to be solved in their conscious relationship. Unlike our forebears, we no longer have the problem of explaining how an "immaterial spirit" like the human *mind*, could relate to *matter* apprehended through sense experience. The differences that we see — what accounts for the need of an individual consciousness to relate to "outside" reality — are *functional*, not metaphysical. Metaphysically speaking, there is no *inside* or *outside*. The knower-known interaction is exclusively due to the different relationships of elements within *matter's energy*. Organic individuals maintain themselves within the wider context of *matter's energy* and have developed consciousness to do so.

The homogeneity of matter's energy dissolves our ancient epistemological dilemmas. For we are finally able to render human consciousness intelligible by grounding it in the sameness of organic reality itself. It reveals Descartes' *res cogitans* to be an illusion derived from an illusion: a redundant exercise within the Platonic fantasy.

the experience of the *ego*

Phenomenologically, the "I" is itself a reflexive discovery of this experience; it is not non-existent, but it is secondary. The reality encountered in the womb by the fully developed human fetus is simply *existence-in-time*; *presence* is global and undifferentiated, without regard to subject or object.[172] But at birth, there is a rude awakening. The human infant immediately begins to *learn* that she is a separate organism, an insatiable center of the energy of *existence* in a world full of similar energy. Her configured portion of matter's energy (her body-person), by making her experience herself as *need, want, desire*, what we have called *the **conatus**,*

---

[171] Cf Heidegger, 1927, 72, *Phenomenology*,p.66: "the cognitive faculty is not the terminal member of a relationship between an external thing and an internal subject; rather its essence is the relating itself ... *Dasein* (the human being)... is always already immediately dwelling among things."
[172] Gordon Alport, *Becoming*, 1955, Yale U., pp 42 and 44. *"The young infant has ... in all probability no sense of self identity."* This may have been a "guess" in 1955. cf, Michael Greenberg "What Babies Know and We Don't" *NY Review of Books*, March 11, 2010, p.27. "We know that children's conception of a continuous separate self develops slowly in the first five years." Review of Alison Gopnik, *The Philosophical Baby*, Farrar, Straus and Giroux, 2010.

proclaims that the primary thrust of her human *presence* is energy and passion for survival. The "I" experiences itself as a point of need because it is immediately assaulted by an avalanche of internal pressures focused on living. Our little sister's perception of reality quickly organizes itself into subject-object driven by the relentless insistence of the subject pole (the *ego-conatus*) to *survive*. Her first vehement cry as an infant announces clearly that she is *beginning to learn* that she is an individual self — a throbbing knot of *being-here and* an unquenchable thirst for *staying here*.

The infant's early experience, which is at the base of her learning who she is, is recapitulated, hopefully with less trauma, in every later experience of conscious life. Sensations bombarding the organism are the undistinguished dual reality of knower and known, the raw flow of *existence-in-time*. The presence of consciousness, however, reveals itself immediately as a *center* in this mix by selecting and classifying the incoming perceptions under the pressure of *self-interest* for the preservation of *existence*. These later experiences are all built on earlier learning. And the long-term process of discovering and identifying herself continues into maturity and beyond. Please note the structured sequence here: consciousness, driven by need, discerns itself as individual *within* a homogeneous totality. But its initial field of perception — experience — is of the totality.

So *being-here-now* for the human individual is simultaneously a drive and a perception, a "seeing" that is driven by vital organic self-interest, the *conatus sese conservandi*. The organic integrity of the human being is revealed in this perception. For the notions, "seeing" and "wanting," are not discrete experiences, neither in kind nor in sequence. We distinguish them only artificially. There is no empirical (experienced or scientifically observed) separation between mind and body, seeing and wanting. It is *the fire of the hunger to live that lights up the world through the eyes of consciousness.* Consciousness is life's tool, its agent, its servant. If there is a priority here, it is to be found in the directing stewardship of *matter's energy* that has placed an irrepressible hunger to survive in the organism. Being-*here-now* is intrinsically orientated toward *staying-here-endlessly*. *Matter's energy* has only one abiding interest: *to exist.*

Learning (time-based *a posteriori* concept formation) *only incrementally* produces the awareness of even the most fundamental objects. **The** *concept of the "self"* is formed no differently. The experience of the self is the result of a *learning process.*

A concrete example: We can all agree that permanent psychological damage will occur if an infant's early contacts with reality are *malevolent.* The ego-unclarity of infancy means minimally, in these cases of severe abuse, the internalization of ambivalence about oneself and, later on, its damaging effects, persist. The phenomenon is a clear indication of the *undifferentiation* with which reality is initially embraced. The infant cannot distinguish herself from her environment, and in fact, even under conditions of normal development, the individual remains vulnerable to this potential confusion for many years, even into adulthood. That the person builds up final interpretations of very fundamental questions, like the identification of the self, only after multiple "insights" preceding maturity, must be taken *as constitutive.*[173] *The accumulation of experiences is an essential part of the formation of the concept of the self.* The sad fact that in adulthood a serious regression still remains possible, shows that even something as fundamental as the awareness of "self" is not a permanently fixed feature of consciousness. It emphasizes the primary status of the substrate for consciousness and the ongoing, secondary, constructive nature of interaction between knower and known. *We are ourselves a continuing process* in which the *presence* and the character of the "self" is discovered and continuously evaluated. The "self," like any other concept, is a tentative, time-related, *a posteriori* label aggregated from myriads of experiences distilled from the totality and open to "reclassification." It is the result of a *learning process.*

This is not insignificant. The reflexive awareness of the "I" is not only *not* the result of an *a priori* intuition, but also it is *not* the result of a *positing*; it's not done on the initiative of the subject, and it's not an *instantaneous*, fixed notion. There is no direct, intuitive perception of the "self." The "self" is discovered and learned (interpreted and conceptualized) like every other object of knowledge — based on concrete evidence and the repetition of interactive experience. It is entirely *a posteriori.* The self is a developing concept, a label created by the repeated experiences of the driving urgency of need and desire, the *conatus-in-action* arising from the individual human organism formed in the process of surviving in an indiffe-

---

[173] Greenberg, *op.cit.* p27; cf. Harry Stack Sullivan calls personality a "hypothetical entity" produced by social interactions. Hendrik Ruitenbeek, ed.,*Varieties of Personality Theory*, NY E.P.Dutton, 1964, p.122

rent world. The awareness of the *ego* is not an action, but *a passion*. It is *generated* and sustained by a *lust for life* that, because it is common to all living things, we know emerges from the organism's elemental substrate — *matter's energy*. It is beyond conscious control. It establishes the ultimate horizon and therefore the principal characteristic of *being-here*. It is an *energy* — the primordial energy of matter — and we are not the authors of it, even though *it is what we are* and defines everything we do.

human intelligence

In the case we are considering here, *human intelligence* with its capacity for generating verbalized concepts, emerged from an animal intelligence adapting to the conditions that characterized the Pleistocene Epoch. The adapting species of *homo,* coupled with the fauna and flora, atmosphere and climate, plate tectonics and sun-spot cycles and most importantly, humanly chosen social structures, language, teaching techniques and cherished values of the last 2.4 million years can be said to have *collectively produced* the sub-species *homo sapiens with* the capacity to use verbal conceptualization[174] Any "philosophical" determination of what it is to be *human* must take this entire collective creative process into account. Our current power of word-managed abstractive consciousness cannot be examined and judged out of context, as if it had emerged full blown, or integrated into another world. Human consciousness is a product of the entire earth-matrix community in an evolutionary process through time.

3. animal survival, evolution and human intelligence

For the human being, consciousness is not at first a "seeing" of anything. As we've been saying, it is part of an animal organism's engine of life. Because it is all about survival, *staying-here* is the very point of this relationship for which abstractive conceptualization, and the *knowledge* that results from it, is so central. The perception in which the subject pole awakens is focused on living and staying alive. Knowledge is not at root seeing but living — driven by a frenzy for *being-here.*

---

[174] Peter Richerson and Robert Boyd, "The Pleistocene and the Origins of Human Culture: Built for Speed," Version 1.1. February, 1998. For presentation at *5th Biannual Symposium on the Science of Behavior: Behavior, Evolution, and Culture.* February 1998, University of Guadalajara, Mexico.

As far back as paleontology can trace, our *animal* ancestors had been doing, not just something similar, but *precisely* the same thing. Human intelligence did not awaken in a torpid lifeless darkness, but in the full brilliance of raucous day. The world was teeming with animal life. The very first stirrings of specifically human perception were themselves **necessarily imperceptible as human** because they appeared amidst the animals, immersed in exactly the same frantic tasks, the same contentious goals, the same terrifying dangers, in the same community of loving familiars using the same practical solutions as they had for millions of years before *as animal consciousness*. The very first "human" was necessarily born of an animal in an animal community, and if she survived it was because she was fully integrated into their animal processes of survival and *communicated* with her non-human parents, siblings and companions about them. Animal survival was, necessarily, the totality of her world. She would have had no focus that she was not directed to by the animals. There was no difference in kind or goal, intention or attention, action or passion. The only difference was *degree*. The newly emerging human did only and always what her animal parents and siblings did, except she did it slightly faster, with a little less effort, fewer repetitions, fewer mistakes. She was better at what it was all about — surviving — and so she survived and her progeny prevailed, and later grew and developed into something noticeably different from their forebears. But the change was, at first, *necessarily imperceptible*.

The fossil record indicates that incremental development was repeated over and over again in human pre-history, eventually producing our genetically specific *variation* of human being, the sub-species known as *homo sapiens*. Paleo-anthropologists have identified at least 7 distinct sub-species of evolving *homo*, our species, that have existed over the last 2.5 million years.[175] Six of them, of course, are now extinct. *Homo sapiens*, in fact, is a sub-species of *homo* and a rather late-comer, having been-here for less than 200,000 years. (This is without mentioning the multiple species of even more primitive hominids, like the various *australopithecines*, who separated from the chimpanzees 6 or 7 million years ago and preceded *homo*.) But the 7 known variations of the later species *homo*, indisputably human, have been identified by significant differences in cranial size, tool use, social complexity, and interest in transcendence as shown

---

[175] Dougal Dixon et al., *Atlas of Life on Earth*, Oxford, Andromeda Books 2001, p.318f.

by evidence of rituality — adornments, burial practices, cave paintings, etc. Unfortunately, none of these other sub-species has survived, leaving us with little knowledge of their cognitive or language abilities. We also do not know whether or not they may have developed in ways similar to our own. After all, we ourselves, upon emergence, were very different from what we are today, perhaps not anatomically but certainly culturally. The primitiveness of an initial phase does not necessarily define the extent of future development; that is obviously true in our case. Yet ours is the only "case" we will ever know about. We have no idea what the "Neandertals," whose brain-size, by the way, was larger than ours, might have achieved (and evolved into) had they survived.

It has been further *assumed* that our version of humanity was organically *superior* to all the other species of *homo*. That we were what we like to call *smarter* is another assumption associated with a theory of "linear" development that is supposed to explain why none of the others survived. But, in fact, at least one of these sub-species of *homo,* the Neandertals, was a *parallel development* to us in time and location. Our particular version of *homo* may have had an intelligence that was just better at eliminating competitors.[176] Perhaps it was because *we "sapiens" were just more paranoid and more lethal.*

We were predators. It seems plausible, according to evolutionary anthropologists, that the *eating of meat* had a crucial influence on the direction taken by human evolution.[177] There was a readjustment of the size and

---

[176] Theories on the extinction of the Neandertals were based on mtDNA evidence that indicated that while Neandertals co-existed locally with *homo sapiens* there was apparently no genetic exchange. The theory concluded that Neandertal disappearance *was not due to absorption* and given the adaptability of all species of *homo*, competition for scarce single-niche food sources could also be reasonably ruled out as an explanation. This would leave only extermination. Cf. Ian Tattersall, "Once We Were Not Alone" *Scientific American Special Edition*, June 2003, Volume 13, Number 2, Pages: 20-27. This theory has very recently (March 2010) been called into question by new evidence that some genomes of *homo sapiens* contain Neandertal sigments. Cf., http://en.wikipedia.org/wiki/Neanderthal

[177] Edward O. Wilson, *on Human Nature*, Harvard U.Press, 1978, p.93 says that "*Early human beings filled an ecological niche: they were the carnivorous primates of the African Plains.*"; cf Gary Stix, "Homo Carnivorous," *Scientific American*, June 2004 p.25Bf. Stix suggests there might be other characteristics of our humanity that developed as a result of being meat-eaters and hunters of animals. Among them he lists the extended period of maturation, made necessary for apprenticeship in the hunt, the longer life-span, and stone tool use for butchering carcasses resulting in smaller mandibles which allowed for greater brain size. The mandible mutation is not a conjecture. It is a fact of the fossil record. It occurred 2.4 million years ago and was instrumental in the evolutionary divergence of *homo* from other hominids.

location of the mandibles about 2.4 million years ago due to the eating of soft food, like flesh, that allowed for an expansion of the cranium. This correlates with the beliefs of paleontologists, supported by associated evidence, that the "gracile Australopithecus," our ancestor, was a hunter.[178]

So conceivably, an enhanced psychological *capacity for killing*, the adjunct to a predatory life style, may have been the inherited feature of our conscious apparatus that brought *homo sapiens* into ascendancy among our sibling species. A mega-millennial immersion in the world of predatory activity would be likely to generate an appropriate *defensive* instinct, a wariness about the intentions of others, as well as the ability to extract ourselves affectively *from the present moment*. We had the consciousness of predators. It gave us a unique *emotional distance* from the animals that we killed or who wanted to eat us. It was a survival-based partialization that falsifies *existence.*

Social benevolence (altruism) and its associated contemplative cognition, an instinctive feature of reality derived from the communitarian nature of *material energy,* is not necessarily rewarded in a world of survival and therefore cannot be identified as the point of the evolutionary lance.

The evolution of human intelligence is a function of survival, not of a transcendent consciousness focused on the "Beatific Vision." Nevertheless, how we subsequently *decide to use* the apparatus that survival has evolved, while possibly extraneous or tangential to its original evolutionary "purpose," is a perfectly authentic exercise of human choice. It's what some evolutionists call an *exaptative* development:[179] a trait evolved for one purpose that gets used for another.

Why these other species of *homo* have not survived, therefore, is not irrelevant for the ultimate interest of our enquiry. Nothing says that parallel species cannot co-exist. There are at least four parallel species of apes — chimps, gorillas, orangutans and bonobos. Why couldn't parallel species of *homo* have co-existed as well?[180] Is it pure fantasy to suggest that *homo sapiens*, with its new "raptor's claw" intelligence was no more inclined to peaceful co-existence with its sister sub-species than any other

---

[178] Barbara King, Ph.D. lectures on Biological Anthropology, The Learning Co. 1999

[179] Stephen Jay Gould, *The Structure of Evolutionary Theory*, Belknap, Cambridge, 2002, p.1232f.

[180] The very discovery of evolution in the 19th century depended upon the simultaneous presence on earth of various life-forms whose step-wise similarities stimulated insight into their genealogical relationship. The historical record was available in living display. It's odd that that should not be true in the case of humankind.

carnivorous predator? The local clans of *homo sapiens* could be expected to kill (and eat) any and all rivals, including other *homo sapiens*.[181] Defenseless against this superior "brain" power, all other sub-species would eventually be exterminated. *Homo sapiens*, however, was guaranteed to survive; for no matter how much slaughter goes on among *homo sapiens* themselves, the sub-species survives because the victor is always a *homo sapiens*. Thus the disappearance of all other sub-species of *homo* might be said to be attributable to the peculiar character of *homo sapiens*, and not to the "unseen hand" of evolutionary "progress."

> "... man is the only species in the animal kingdom that will perform wholesale massacres on its own members; animals are protected from doing this by innate behavioral control of aggression. Perhaps it is because man is a recent species, lately descended from australopithecines who owed their survival to aggressive behavior in bands with all that implies in the way of mob psychology."[182]

conclusions

Process thinking affects all aspects of the once settled questions of philosophical enquiry. Take the essentialist theory of exclusively **human "spirituality:"** the theory rests on two ancient assumptions: one, that there is a single universal definition of humanity through all of evolutionary development, and two, that there was a "clean break," an unmistakable difference in kind (and not only in degree) between animal and human consciousness when it emerged. Evolution has demolished both those assumptions. For it seems utterly absurd to claim that the conscious operations of the entire variegated spectrum of evolving humanity, from the earliest *australopithecine* hominids through the current model of *homo sapiens*, were always and everywhere the same ... or that the earliest sub-species of *homo* could be clearly distinguished from what was animal. The facts seem to indicate the opposite. There were, rather, a *continuous sequence of incremental changes* without there being, for a very long time, any way of differentiating human from animal consciousness. Even today, the affinities between humans and other primates surprise us. How much greater must have been the similarities between the

---

[181] It has been suggested that the human attraction to pork meat is attributable to the affinity of pork meat and human flesh.

[182] De Beer, *EB op.cit.p.21* ... also Cf William McNeil, *The NY Review of Books*, April 17, 2008, p.48, reviewing a book on Genocide, McNeil writes, "Our earliest ancestors probably extinguished competing hominids so *Homo sapiens* alone survives."

animals and the more primitive varieties of *homo* that no longer exist? This renders the theory of exclusively human spirituality a projection.

It tends to corroborate that there is no such separate super-genus of being called "spirit," and, what we have called "spirit" and identified exclusively with our human rationality is, in fact, the expression of a conscious relational *energy* that pervades the entire universe in a *continuum* of gradations from the foundational "strings" to the highest forms of life as they now exist. We may identify this force as the actual *material energy* of the universe, expressing its inner vitality endlessly in the *present moment*, evolving according to the conditions in which it finds itself. This energy is the *physical presence* we experience, what I am now calling *existence*. The sensation of *existence*, the experience of *being-here-now*, is the energy at the core of matter and all degrees of consciousness.

In such a context, human "transcendence" becomes self-contradictory. Human abilities once claimed to go "beyond" the capacities of "mere matter" are now seen to be elaborations of the very properties of matter itself. *Matter's energy never transcends itself* — its potential — even as it transcends the forms that currently exist. The relatedness within *matter's energy* is universal; it embraces absolutely everything and is another aspect of the radical unity of *existence* from which communitarianism arises. Human consciousness, whatever its as yet unknown capacities may turn out to be, is in unbroken continuity with the entire dynamic Universe.

We ourselves are this matter-energy *and always have been*. Once we realize there is no independently existing "spirit," it should finally dawn on us that *we are* the very same *material energy*. We are *existence* and lashed forever by an inner compulsion to continue to *be-here*. On the personal, conscious plane it takes the form of *the drive to survive*. It is not a choice for us because *matter's energy — existence —* is not a choice for us. *We are existence and we cannot NOT be existence:* **we have to exist.** That we hunger and thirst for it is no mystery to us at all. *We want to be here.* And in this sense we *understand* it intimately, even though **we do not know** what it is.

Our insatiable thirst for immortality and our projections of an afterlife, come from this as does our characteristic obliviousness of death. Despite living with it everyday, when the fact of death comes home in an event of unshielded *realization,* it is a shock that has the capacity to immobilize us. We cannot imagine *not being-here*, and if we cannot be-here in reality, we will fantasize it shamelessly. We deny death to its very face. Nothing can stop us from wanting to *be-here* endlessly ... no matter how incontrovertible the evidence that we will not. Such is the **conatus** we were born with, the *existential dynamism of material energy*.

understanding and knowledge

Our theory of human abstraction, presented in chapter 5, introduced us to a wider perspective on consciousness. That chapter concluded that there is nothing immaterial or "other worldly" about abstraction. Abstraction is an instrument of human consciousness focused on survival. This present chapter examined the organic nature of human relatedness and its evolutionary development. Together these two chapters should have put to rest the dualist theories that tried to tear us from our organic matrix, projecting a hostile relationship between an imagined immaterial mind and a body made of dumb, inert, passive matter.

Consciousness is an adaptive response to the homogeneity of *matter's energy*. It provides **experience,** the relational side *of the self-embrace of matter's energy*, which is the central dynamism in our universe. Humanity evolved as a product of the survival efficiency of its species-specific conscious abilities. Language-based abstractive conceptualization, as analyzed in chapter 5, is exclusive to humankind. This specifically human variation of animal consciousness — the use of word labels — has made the human phenomenon, with its constructions and destructions, possible.

In these reflections the broadest possible take on the human version of consciousness will be called *"**understanding.**"* *Understanding*, then, as used here, does not necessarily correspond to the end product of word-labeled abstraction, a term synonymous with "knowledge." *Understanding is broader than knowledge. **It includes the body**, occurs even in the absence of words,* and therefore is not easily objectified. Knowledge is subordinate to and functions in the service of understanding. Knowledge is only one route to understanding.

The following chapters will examine some of the cognitive operations that exemplify an *understanding that bypasses knowledge*. They will illustrate the amount and quality of organic conscious comprehension that we humans have at our disposal. Our human intelligence is a rich and varied tool-kit that defines our immersion as conscious organisms in the community of *material energy*.

chapter 7
co-existent contemplation

We have been insisting on the priority of the survival instinct in the evolution of the specific capabilities of human consciousness. We have been saying that the demands of survival explain the unique abstractive operations of our intelligence. We even claim that the survival drive is the origin of our awareness of our very selves, the "transcendence of the *ego.*" We have understood the instinct to survive to be an expression of the dynamism of *matter's energy* itself, the building blocks of all reality. This understanding places an existential *hunger,* a *conatus,* something of a selfish frenzy at the very center of human life as it does for all life.

But there seems to be something missing here. For we all know quite well that our cognitive functions do *not need to* respond to such "selfish" impulses. The disciplined procedures of scientific enquiry, for example, are committed to the avoidance of self-interest, and most certainly would deny any role to frenzy. In this chapter, I'd like to reflect on a phenomenon that seems to contradict our claims for the defining role of the survival dynamism. I call it *co-existent contemplation.* The term refers to our ability to relate to *existing things* with a quiet recognition that is beyond the distracting demands of self-interest or need.[183] Co-existent contemplation appears to be a simple "seeing." But if consciousness is organically structured as a survival tool, how is that explained?

For ancient Greeks like Plato, the serenity of contemplation was a seminal and revelatory phenomenon. Contemplative comprehension was believed to be the apex of human achievement, a natural activity of the

---

[183] Though I was unaware of Schopenhauer's treatment of "aesthetic contemplation" when I was writing these reflections, it should be acknowledged that he develops his analysis in response to a similar anomaly, for his "Will" in many respects corresponds to what I have been calling the *conatus,* the "drive to survive." The analyses are remarkably parallel, for better or worse. Given the established primacies in each case, "contemplation" requires an explanation. See the appropriate Appendix for further discussion of my similarity (and dissimilarity) with Schopenhauer. For the similarity with the question of altruism see chapter 12.

human mind which manifested the presence of immaterial "spirit." Cognitive serenity, in this view, which included a dispassionate and undistracted objectivity, was thought to be the natural and original condition of the exercise of human intelligence. The Greeks connected contemplation with abstraction which they believed was a divine power of insight that generated universal and necessarily "spiritual" ideas. The intense selfishness that actually characterizes the human condition, Platonists explained, was an anomaly, an unnatural state of affairs — in fact, they believed it was the result of a moral collapse, *a fall*. Our current way of thinking and acting is *pathological* and requires corrective action. Therapeutic efforts made in this regard, if sustained and practiced well, they said, would return us once again to the pristine contemplative serenity which is our nature. The similarity of this programmatic dimension of Platonism with the essential dynamics of Christian "redemption" should not be lost on us. Christianity, through Philo of Alexandria, assimilated to Platonism on this question. The "fall" for Christians was "Original Sin."

By condemning as unnatural the fact that we are *not* born disinterested and detached, Plato was forced to have recourse to this imaginary scenarios to explain human need and craving — which for him were more than a problem, they were an intellectual paradox. The loss of primordial detachment and peace-of-mind, he said, must have been due to an event that affected everyone, because we are all born pathologically "hungry." Plato imagined Spirit "fell" and was *punished* by being imprisoned in matter, and thus *lost* its dispassionate contemplative objectivity.[184] We must remember that in Plato's world of matter and spirit, peaceful contemplation was not possible to matter, and spirit was believed to be beyond the stab of hunger.

There was a variation on this theory of a *fall* from nature elaborated by the architects of Greco-Roman Christianity, Augustine of Hippo in particular. It identified Plato's "fall" with the Genesis Myth of the Jewish Scriptures. In the Christian version, a "spirit-dominated" humanity was fatally corrupted by Adam's sin resulting in a *metaphysical reversal* of the natural order. Human beings lost an original *immortality* and came to be unduly influenced by their animal body with its inordinate hunger due to *matter*. While this effect was *unnatural*, it was true for each person from birth. In effect, they said, the human spirit had become a prisoner of the flesh.

---

[184] Plato, *Phaedrus*, Jowett tr, *The Works of Plato*, NY Tudor, p. 403 ff.

There are certain features of this version that Christians consider important to their ideology, which they claim makes it different from Plato's. But from my point of view the two versions are fundamentally the same. For they both declare the real world *unreal,* in the sense that it is *not supposed to be the way it is.* Their gratuitous projections about a mythic past resulted in a split universe comprised of two different kinds of "being," *an imagined spirit and a corrupt matter.* They assign all vitality, energy, dynamism, rectitude and equanimity to spirit, while declaring matter dead and inert, an absolute emptiness and thus the source of need and craving. As opposed to spirit, matter lacks "being," is therefore hungry, selfish and grasping. Spirit is self-possessed, objective, serene and emotionally dispassionate.

a different perspective

In the ancient hypothesis, of course, the emotional detachment which characterizes contemplative consciousness also implied the human ability to transcend the *immersion-relationship in matter's energy.* The human spirit, in other words, was believed to transcend "matter." That also meant that Plato's vision accepted a sharp differentiation in knowledge between subject and object. For the subject was a "spirit," and believed to be above and beyond the world of the object, "matter." It knew itself as a "self," i.e., a "spirit," by a direct immaterial intuition, and hence it knew "matter," even its own body, as "other than itself."

But in the view espoused here, this is not the case because we know that the relationships within *matter's energy* form the foundational horizon for human consciousness. Human intelligence not only was evolved *from and by* material energy, human consciousness *is matter's energy* in its (up to now) most developed form. So the global self-embrace of *existence* is more primordial than either the perception of self or the specific activity of human intelligence making selections within the undifferentiated welter of incoming data. The relationship to *existence* is *prior to the subject-object division* and therefore it is more basic than abstraction.

The perspective proposed here can be seen to differ radically from the ancient in this: *the immersion in existence necessarily implies the perpetual standing perception of the global continuum of existence as prior horizon. Matter's energy (existence),* the substrate of the totality of things composed of it, is the source-matrix of human consciousness; it is "earlier" than word-concept selection. Abstractive conceptualization must radically modify the more primitive un-differentiation if it is to achieve its end. The homogeneity of *existence*, therefore, necessarily remains a most fundamental perception, irreducible and un-suppressible, always and everywhere in operation, a standing condition, the horizon within which the survival drive and its associated abstractive selections function. It is a background that must always be there. Therefore attention and focus may always turn to it.

This forms the basis for understanding the subject matter of this chapter. Co-existent contemplation is a conscious perception that represents a disciplined *intentional* return to the global un-differentiation characteristic of the pre-abstractive relationship of human consciousness. This unselected background perception is epistemically *prior to* the developments of organismic evolution which produced human verbalized conceptualization. Therefore it is something we share with the animals who are also in conscious contact with the material world around them .

The background horizon, however, the global data from which selections are made, is continuously apprehended as *present* to consciousness though not necessarily in focus. The ability simply to contemplate, without need, derives from the irrepressible organic core of human conscious presence, the constitutive immersion relationship in *matter's energy*. Because of this constitutive "pre-evolved" relationship, all consciousness, human or otherwise, in embracing *existence* in any form or manifestation, is embracing itself. The fundamental global receptivity of consciousness that precedes and is presupposed by the act of selective abstraction and conceptualization is the way this shared inter-species relationship is manifest in human beings.

I don't want to give the impression that there are two separate operations or "objects" here, the one "background" and the other "foreground;" it's more integral than that. The "background" apprehension of all of what's there in any given moment is the entire field of data within which consciousness makes its selections. What selects for survival, *the subject*, driven by the organic basis of the surviving self, is an intrinsically connected and related part of the whole; and its movement cannot be considered apart from the global field of perceptions within which it selects and with which it engages continuously. But at any point, stimulated by some new interest or another, *consciousness may purposely turn its attention to the entire field as field*, or some objects or even the very same object within the field but also *as field*, that is, apprehend it *precisely in its quality as background.* So, even though selected and brought into focus, it can be viewed in its pristine condition as non-selected, *not the object of survival interest*, and therefore relatively undistorted by human need. (The new focus of interest, admittedly, may subsequently skew selection in its own way. It may even present itself as necessary for survival under a new rubric, perhaps as social or peer pressure, or perhaps eternal salvation and lose its disinterestedness.) Thus, a consciousness whose *interest* in what is spontaneously selected for survival has been suspended, can utilize its *special* abstractive apparatus to embrace perceived objects in an act of co-existent contemplation.

the oneness of *existence*

There has to be a concrete homogeneity, a common element, shared between knower and known that makes such an organic relationship possible. This common denominator is *material energy* itself, *existence*, which we are claiming is the energy which is matter, the "stuff" of all things including us in every respect — "mind" and body, flesh and "spirit," and, of course, consciousness. Material energy is all we are. It is all *everything is*. It is both knower and known. Consciousness, broadly speaking, embraces all relationships within *matter's energy*. The *species specific* human variety of consciousness includes our abstractive survival apparatus, but the general conscious relationship to reality we share with all things according to the measure of our respective capacities. The cognitive side of this *general relationship* is our co-existent contemplation.

So our capacity for the serene appreciation of the entities within *being-here*, to which we relate without regard to our needs, is a display of the material unity at the base of reality, the identity of all of what is *present*. A corollary, therefore, is that it is always insuperably physical, for it could not be uninterruptedly present to our physical mental apparatus if it were not. This material homogeneity, shared by knower and known, is exactly the physical, observable, measurable existent *energy* at the base of all things, what scientific theorists say might be *vibrating strands* called "strings" responsible for the existence of sub-atomic particles — quarks, gluons, electrons, neutrinos — the building blocks of everything we can know. *Existence-in-time* is one homogeneous physical phenomenon. That is the basis for our ability to apprehend it, to survive within it and to contemplate and appreciate it for what it is.

metaphor and distance

The objects of contemplation differ significantly from the objects of the selective operation of abstraction, not because contemplation is a different faculty or a different operation — it is also a selection — but because contemplation focuses on an "earlier" stage of conscious perception; it grasps reality "ahead of" the verbalized selections driven by survival needs.

Because contemplation looks at the background in its pristine, "un-abstracted" condition, it may be forced to use tools of expression that differ from those offered by warehoused word-labeled normally interested conceptualizations. It will, minimally, have to focus perception *"before"* *society's word* and from there decide what *form of expression* can be trusted faithfully to represent the new "disinterested" perception. Perhaps a new word will be needed. We will explore this in the chapter on *metaphor*.

Contemplation involves a *de-emphasis on the distance* separating knower from known. We remember, of course, that the subject-object distinction, which constitutes the "transcendence of the *ego*," was itself a *learned* reality, secondary to global perception and impelled by the *conatus,* the organic drive to survive. So, according to our hypothesis, it is the *hunger* for survival that is the origin of the awareness of the "self." Otherwise the objects of perception would retain a neutrality which is the residue of their original un-differentiated condition, the homogeneity of *existence.*

Schematically, then, two parallel lists of features correspond to these two *foci* of consciousness. In the first, *abstractive conceptualization* is tied to the *separate ego,* spawned by the *drive to survive*, resulting in the thing known being relegated to the subordinate *status of "object,"* judged solely from the point of view of its *benefit* for the knowing subject, identified, labeled, evaluated and stored by the community in *conventional language-concepts.*

In the other, *contemplation* implies the polar opposite. It tends to create a *"new word,"* a *metaphor*. It imposes a *quiescence on the hunger* that drives the abstractive process. It entails the *suspension of selectivity* based on need, the *subsidence of the sense of self*, the *muting of differentiation* and the distance between subject and object. It requires the *lessening of the community's role* in the interpretation of experience.

Contemplation is a global embrace. Hence, it signifies *co-existence* for it does not subordinate one entity to another. It accepts all as *co-present*; as *being* equally *here in the present moment.*

contemplation a product of community

We have emphasized that contemplation focuses on a *background* perception. This is important to our point of view. For the evolved dynamism of consciousness, special to humanity, is *not* spontaneously driven to co-exist and contemplate but rather to abstract: to select, conceptualize and survive. The contemplative focus by itself would not have been able to survive in a predatory world, and therefore could not have been the cutting-edge of evolutionary development. This provides the ultimate contradiction to Plato's theory of the natural primacy of contemplation.

Clearly, we are radically *capable* of directing our attention and intention as we choose, but it seems that we can only afford to do so when the intense pressure to survive has been (temporarily) suspended. Only the freedom from want or danger *provided by human community* can create the conditions necessary for contemplative cognition.

So we encounter an anomaly. Human community is necessary, and must necessarily function providing respite from the lash of survival for there to be the conditions present for contemplative cognition. But ironically it is conceptual language, the very tool and repository of society, that has already focused the *meaning* of potential perceptions in the biased direction of human survival. It seems that language, therefore, against its normal inclinations, must begin to include the *meanings* of words developed *outside of the conventional mainstream* if it is to provide support for the contemplative function. Only *metaphor* as a praeter-conventional perception can create these new meanings, layering onto language the various significations that contemplation discovers. Society must be capable of incorporating the contemplative meanings newly grounded by metaphor if it is to provide sustenance to this most precious and cherished operation of human intelligence. This presages the exponential expansion of vocabulary and grammatical expressability which will make language into the rich and flexible tool we know and cherish.

creativity from stress?

There is another derivative of the essential role of the community in "permitting" or "fostering" contemplation ... and that is the age old tendency for the powerful artificially to create conditions for the pursuit of contemplation by enslaving others, who are forced to work in their stead, leaving them free for "thinking" as a leisured pastime.

In this regard I will make one quick side commentary on contemporary opinion and social practice. Our current "market-driven" economic philosophy proposes, as a matter of principle, to re-insert the individual back into a struggle for survival that presumably characterized our earliest societies, *purposely in order to* stimulate "motivation" and "creativity." We must realize that a society under such a regimen would deny freedom from worry to all but the leisure-class. This "whip" economy intends to mimic the Pleistocene jungles where our consciousness emerged. Admittedly most of us still have the capacity for that kind of life. It's the way our consciousness is organically structured. But it must also be recognized that the elimination of the economic security that only society can provide and the return to more primitive conditions also militates against the possibility of facing larger environmental issues that are dependent on disinterested appreciation.

At the risk of sounding apocalyptic, I would emphasize that the collective destiny of all life on the planet earth depends, at this point in time, not on the individual energy of a human predator focused on his prey and those who would prey on him, but rather on the development of a dispassionate understanding that can look at what impacts the survival not only of the human species, but of every life-form on the planet. If we continue to encourage self-interested, *non-contemplative perception* by reproducing the stressful survival conditions of the prehistoric savannahs, where predation and accumulation were necessary to life, we will maintain habits of consumption that are currently pushing the resources of the planet and its capacity for regeneration to the point of no return. The very dynamics that evolved us into predatory little paleo-primates, if continued in the form they had when we evolved, will destroy us. We must begin to think *contemplatively*, disinterestedly, co-existentially. The time of individual competition and accumulation is over. This is a confirmation of the discovery of chapter 6, that the individual is depends on a larger community to provide the security that allows for contemplative understanding. We are no longer predators. To run economic life as if we were, is an anachronism we cannot afford.

Society must provide the "leisured" context, the economic security for this contemplative perspective. But contemplative awareness can function *only if we choose* to bring it forward and to make it the focus of consciousness. The contemplative function is founded on our ability to focus intentionally on the background. Contemplative regard and its accompanying attitude of co-existence is first generated and then sustained *only by choice and discipline.* It is a socially grounded and socially sustained operation. It can only be done with the help of the community.

This is important to our understanding of who we are, how we function and what kind of economic system it behooves us to construct. We must recognize: *contemplation is not the spontaneous focus of human intelligence.* It is a capacity that must be supported by the larger human community if it is to be nurtured in the individual. It's another of our socially rooted *learned responses.*

Contemplation is the gift of the community to the individual even as it was originally and continues to be the gift of the creative individual *to* the community. It is *not* the reason why we evolved and therefore *we cannot take the contemplative function for granted.* We evolved as survivors in a predatory world. It is even probable that we were predators ourselves. There is nothing to prevent us from reactivating our predator potential, physically or socially. The Platonic Paradigm, with its separate realities of matter and spirit, stands in utter contradiction to the proper understanding of ourselves and our connection to the organic earth. I submit that in large measure the ecological problems we face at the present time, which have already led to the extinction of life forms other than ours on a mass scale, are fundamentally traceable to having severed ourselves from our balanced place in the earth's network of living things. Our own survival is next. Ironically it will be our ability to transcend our originating structures and to see this connection with the earth through dispassionate eyes that may permit us to survive.

contemplation: the new point of the lance?

Contemplation is something we cherish and love. We ourselves *want* to pursue it, even though it does not define our origins. It is another "exaptive" response. We can begin to see that we may be witnessing the beginnings of a new evolutionary "point of the lance."

Evolution proceeded for eons according to the survival formulas set by natural conditions. But with the appearance of the species *homo* 2.4 million years ago, we entered a new era of evolutionary focus. For since the inception of the all-embracing *human social environment*, the requirements of survival have shifted from *adaptation-to-natural-conditions* like geography, temperature, available water, existent food sources, predators, competitors, etc., to adaptation to a *socially generated environment* in which human preferences increasingly determine the survivability of individual organisms — even those of other species. Even as *matter's energy* continues to respond to the primordial need to survive, it finds itself adjusting its responses to accommodate the new social environments created by human choice. Thus we live in conditions of "social Lamarckism,"[185] the inheritance of *learned* characteristics. The survival drive remains ever the motor force of evolutionary speciation, but human choice based on our "leisured" desires, *changes the environment* and therefore increasingly directs the voyage both for ourselves and other species.

Indications that the earth's ability to support life is not infinite, suggest that, in one direction at least, our way may be barred. For we cannot continue on the path of endless exploitation. The traditional model of human domination of the earth based on assumptions about the "divine rights" of our "spiritual" nature *does not work*. Our immaterial illusions gave us leave to act as if we were the "owners" of material creation with unlimited control.

Co-existent contemplation was misinterpreted by the ancients — they thought it was primordial — and that misinterpretation supported a distorted view of human nature. But once situated where it belongs, this most cherished operation of human intelligence can open us to a future where our survival is dependent on the health of the entire eco-system. It is precisely the ability to think holistically, disinterestedly, *co-existentially* beyond exclusively human interests that will do this. And if this way of looking at things is, paradoxically, necessary to survival, aren't we looking at the future of humanity? For whether or not it portends a physical modification of the human organism and its intelligence, it seems undeniable that *co-existent contemplation* is something we have to learn to do, and do habitually, if we are to survive.

---

[185] See appendix

chapter 8
interpretation

person as process

Highly complex entities, like human persons, or individual higher animals, present a unique challenge to our cognitive powers. Such cognitive "objects" are really *processes*; they are not "objects" at all and so they cannot be "known" in the direct sense of the word. They are subjects. They are centers of continuous change. They are definable only by the conjectured *term* of the direction of change as displayed in their behavior. This direction, or "drift," can only be *interpreted ... guessed at ... not "known."*

Subjects cannot be *known* because they are not objects, neither single nor multiple. They are not even the composites of other objects, static or dynamic; they are not objects either before or after aggregation. They cannot be conceptualized. Persons are *subjects,* meaning the source of a driving, changing intentionality, *a process.* The very reality they present to consciousness (whether their own or another's) is itself always in a movement of conscious relating, which is to say, in a continual state of change due to interactions with other things or persons. There is nothing "fixed" in the makeup of the "subject," except, in a manner of speaking, the unique organic configuration at its base, which, albeit imperceptibly because of the difference in scale, is also changing through time. Change terminates only at death when the "subject" ceases to be present.

This is the phenomenon of personality, the *integrated function,* as experienced both internally (by the person) and externally (by a personal observer). Our examination here does not pretend to discover or confirm a metaphysical basis for personality, though some may see such an implication in the very description. Our intention here is not physical or metaphysical. It is epistemological-phenomenological. I want to describe *interpretation, a process* that goes beyond knowledge.

process

Persons are "known" only by their intentions. And intentions are conjectured (not known) by observing behavior. The interaction between knower and known, which includes the occasions when the "known" is also a knower, presents a changing, open, conscious intentionality that can only be apprehended in action — on the fly — by other subjects *who must necessarily have recourse to their own intentionality* in order to interpret what they see in another. Intentionality is not known directly, by oneself or someone else; it is *inferred* and its direction or import probabilistically determined, *interpreted,* (guessed at*!*) by assessing the drift of its various expressions — where it is headed. It functions similar to *the calculus* which also assesses the direction of a changing set of relations. We say we *understand* someone when we know "where they are coming from," or what they're "getting at." Our interest is not *what* they are but *what they* want: the "goal" or purpose of their actions (or lack of actions). It's the evaluation of *a process.* We get a sense of the immense complexity of this moving, goal-directed conscious evaluation when we realize that having recourse to my own "intentionality" means I am simultaneously trying to assess the drift of *my own process* (since "I" am not a direct "object" of knowledge for myself either) even as I use it to interpret the process of another. Each of us has had the experience of "surprising ourselves" by seeing reactions or attitudes of our own, for better or worse, which we did not expect. We get to know ourselves and others by seeing what we, and they, do ... and eventually getting a sense of what we want. In the sphere of *understanding* ourselves and others, we see *what we do,* and from that infer *what we are.*

When we refer to a "person," whether ourselves or another, it is this active progressive intentionality — purpose in process — that we mean. But it should be noted in passing, that this same interactive assessment characterizes conscious contact among life-forms other than human as well, albeit at different levels. Animals also assess our intentions and one another's intentions just as we assess theirs. We are wary of one another, especially in the wild, and an animal will often wait until our intentions become clear before acting. Intentionality and its interpretation is not an exclusive feature of the human "spirit," *it is a function of all living processes and most emphatically for those with consciousness.* It's not just a feature of human intelligence.

object or *process*?

When I see an individual sleeping, I see a human being as an "object" — static, unchanging, a human organism made up of elements, mechanisms, organs and functions with identifiable characteristics, perhaps a name, whose dynamism as a subjectivity (personality) is not currently active. When that person awakens, however, and their subjectivity becomes engaged with mine, I may still "see" a body showing vital signs, a human organism, with my eyes. But what I "see" with my conscious intelligence is something more — what I call a "person," a subject — a continuously changing kaleidoscope of subjective intention and interaction. I encounter a dynamism that modulates with the interchanges of knowing, being known and knowing back. The static "objective" elements, like the identifiable body and its organs, recede into the background as substrate.

We were introduced to this experience in chapter 3 as the phenomenon of the *integrated function.* The individual is not known as an organism of functioning parts and elements, but as an "identity" experienced independently of the organic structures that support it. The organic substructure is so merged into the background that I have to remind myself of its presence. I do not spontaneously attend to it. What once might have been called an "object" has now become an interactive *complex process* in which *that person's* intentionality *and mine* both weave constitutively into the changes that are occurring in each of us. I realize from my experience of my own subjectivity that the subjectivity I am witnessing is itself the product of a rapid and extensive *guesswork* on the part of the person with whom I am interacting, evaluating me and the intent of my actions and words. So I am guessing about the guesswork of someone who is guessing about me.

The past determinations already made about these intentionalities can be a significant factor in the present evaluation. The past, i.e., someone's past actions, may make us believe we "know" that person. But the past is simply another input. It is not determinative, however weighty and repeated, either for the knower or the known. Similarly, ideological commitments, creeds or party lines that project a relatively closed system of projected intentions, however inflexibly proclaimed, are only another input among many others. *There are no fixed values*, by which I mean "objects" that could be "known" in these experiences of subjectivity; and those that are, seem to be restricted to the organic equipment that all human beings share. They hardly offer what we routinely consider "knowledge" of this particular person.

My *understanding* of the person at this present moment is not the result of "knowledge" and therefore cannot be reduced to the classic Greek model judgment — subject, predicate, object — that lies at the base of the rationalist epistemology. To *understand* a person means to generate *an interpretation of their intentionality* that results from the appraisal of a vast array of complex elements *including myself as guide*, many of which are extrinsic to the situation or relationship, only partially known, not fully understood and most constantly changing. I suspect that there is not one single simple element in the whole welter of incoming data about a person that could be called an "object of knowledge."

My own awareness of myself enters necessarily as a heuristic factor in all these evaluations. The depth and accuracy of my self-experience are major items in the interpretation I come up with. *My interpretation of myself is a constitutive element in my interpretation of others.* When I finally finish my assessment of the person I am trying to understand, it is the result of an incredibly involved *indirect* evaluation that *includes myself.* In the final analysis, however accurate my judgment may turn out to be, **understanding**, in this context, is not the result of **"knowledge."** It's the result of a complex conjecture, which I am calling *interpretation.*

words, process, "knowledge" and *understanding*

Process, of which subjectivity is a sub-set, requires interpretation. Interpretation is conjecture. Interpretation is itself a *process* resulting in *understanding,* not knowledge. These are important distinctions that I believe correlate to other aspects of the human cognitive function. Human knowledge, as I have been proposing, is a product of word-guided abstraction. It refers to concepts — the linguistically labeled aggregates of selected perceptions. And while these concepts are themselves the product of a *process* of gathering repeated perceptions, their ultimate denotation tends to appear as a static "thing" especially since they are linked to an unchanging physical "word." This is how it was possible for concepts to be thought to capture "essences." They are intimately and necessarily associated with their material expression — images — be it word, gesture or symbol, which are simultaneously the received conventional labels of the local community for determined selections of data. On the time scale of a human individual, words are static and resist change. There is no "knowledge" that is not conventional, conceptual, abstract and associated with a particular word; it is the synthesis of a (grammatical) subject and its repeated qualifier(s) (predicates) highlighting common selections and given definitive status by the local language community.

Interpretation is so different from "knowledge" in the sense just described that the two could almost be called distinct operations of consciousness. For it's not that the "objective" organic operators that underlay interpretation are simply being subsumed into the rapidly changing process of interpretation, rather the very *intentionality*, the purpose-in-process which is the real "object" of my cognitive interest, bears only an extrinsic (learned) relationship to the "objective" substrate, the human organism which is its source. In other words, I don't "experience" the person as "object;" neither my own or another's because I am not focused on lips moving, eyes darting, idiosyncratic sounds being formed, much less to neurons firing, neuro-transmitters being released, hormones on the move etc., etc. I have to remind myself that these things are occurring. And before the age of modern science, even that would have been impossible because we did not even know that they existed. The organic ("objective") connection enters only as background into my understanding of the person. *What I directly experience is the assessed intentionality, what we call the "person," the subjectivity, the integrated function.* And the assessment *includes an assessment of myself.*

Unfortunately there are no "words" that are *not* the expression of conventional knowledge, i.e., fixed concepts. "Interpretation" does not have its own terminology, except perhaps when it uses metaphor; so the attempt to communicate interpretation necessarily involves approximations borrowed from conceptually designated "objects." The net result is that in our understanding of things human (the area of greatest interest and importance to us), imprecise, improvised terminology, falsely believed to refer to a "knowledge" that in fact does not exist, is utilized to communicate the results of educated guesswork. This is the status of not only our gossipy personal judgments about one another, but also much of what are known as the behavioral and social sciences, psychotherapeutic practice, not to mention judicial judgment, political analysis, moral opinion, literary criticism, and taste in wine — enterprises whose experts utter their pronouncements authoritatively, definitively and "objectively." Such opinions, the reader may agree, are not "knowledge;" much less are they "objective." But they do constitute *understanding*, and depending on the degree of experienced perspicacity brought to the issue by the "knower," they may be quite accurate and deserving of consideration.

Interpretation is a conscious process that produces *understanding* without the use of the traditionally defined operation of conscious apprehension we call "knowledge."

chapter 9
realization

There is another way of *understanding* that goes beyond "knowledge." I believe it's what we all mean when we use the word "realization." The term refers to a familiar phenomenon that is described as an emotional resonance that sometimes accompanies information.

We've all had the experience that we may have "known" something but we never "realized" its significance, what it "meant," until an event occurred to bring it home to us. The word "realization" in this case not only refers to an emotional aura associated with data; it rather denotes an *understanding,* a level of comprehension that goes beyond knowledge.

An example. We don't "realize" how old the earth is. Bill Bryson in his book, *A Short History of Nearly Everything,* says " If you could fly backwards into the past at the rate of *one year per second,* it would take you a little over three weeks to get back to the beginnings of human life." If we continue his comparison, we would need 7 years to get back to the time of the Dinosaurs and twenty years to reach the Cambrian explosion. And it would take 160 years to arrive at the dawn of life on earth ... one year per second*!*

We've all heard the facts concerning the age of life on earth. Life began about 4 billion years ago. Complex cells appeared about 2 billion years ago; the Cambrian explosion of life about 500 million years ago; the Dinosaurs 160 million years ago; and the species *homo* only 150,000 to 200,000 years ago. To say "four billion," or "160 million," etc., gives all the relevant information. No other way of saying it yields any more factual "knowledge." But, those numbers are really unimaginable; so large that we have nothing to compare them with. The result is that we know the fact, but we don't "realize" what it "means," that is to say, in a very real sense, *we don't understand it*. So, in order to help us understand the time scale involved, often the age of the earth and the stages of the evolution of life are given in terms of a different scale, as in Bryson's example.

In a similar example, an actual trip across the US in a car or bus yields a deeper comprehension of what three thousand miles "really means" than just the use of the words. Knowledge of distances is transposed into "understanding" when accompanied by experience. That is *realization.*

These are examples of quantity; but the phenomenon occurs throughout our cognitive experience — in qualitative examples as well. Our love for an individual is "known" and taken for granted until death or a near-death event "brings home" how much that person really means to us. Forces of nature like hurricanes and tornados are "known" to be powerful, overwhelming and terrifying; but until they are actually experienced, what "powerful, etc.," *means* is not really understood — it is not "realized."

When we examine the difference between what I am calling "knowledge" and *understanding,* in these cases we can see that while the facts remain the same, the resulting new comprehension, the "realization," is experienced as a cognitive apprehension that seems to include a new dimension that goes beyond "knowledge." The personal experience, the actual conscious *existence* of the organic individual in the concrete present moment seems to expand the "data-input" *qualitatively* until it crosses a threshold that we recognize, from experience, to be the dawn of a new level of awareness. But we know there is no new data. Is there anything new to account for this? There is an associated emotional component that can probably be measured; but it seems to *result from* rather than produce the experience of understanding. A "dawning," a "realization," precedes and produces the emotion. What is this realization?

a new relationship

First, as we've noted, we know it cannot be called *new knowledge* because we have already determined that there is no new information. So from the point of view of the classic epistemological paradigm of subject-copula-object, there is no new *object.* Certainly, then, we would also have to say there is no new *subject* either, since the person who first knows and then only later "realizes" the significance of what's known, is the same. The answer must lie in the copula which can only mean that the *relationship* between subject and object changes in a way that *transcends* the raw knowledge, the "facts," while simultaneously remaining firmly within their parameters. It is this newly perceived relationship between subject and object, this *realization,* that intensifies "knowledge" and expands it into *understanding.*

The relationship of subject to object is already given in knowledge and is therefore fixed and determined. If there is a new understanding based on an expansion of the relationship of subject to object, the new relationship must be determined, structured, by something other than knowing (and its abstracted concepts) or the already present data. The relationship itself has to be somehow "defined" in terms other than abstractive knowledge and its conceptual products. That relationship is obviously fully conscious, and even, we might suggest, *hyper-conscious* because what we've been calling *realization* results in an *intensification of consciousness.* In a very real sense, as we often say, we are aware that we *never fully understood* the facts until we "realized" what they meant. So minimally we can say that since there is ultimately produced an *understanding* that goes beyond the limits set by verbalized abstractive conceptualization (knowledge), it must have been due to a new conscious perception that goes beyond those that constitute conceptualization.

the shared organic dimension

I believe that this new relationship between subject and object includes an awareness in the subject (the "knower") expanded by precisely those dimensions that are *homogeneous with the object completely apart from specifically human cognition.* This is the key. **There is an organicity shared between subject and object, matter's energy, not revealed in conceptualization,** in which their shared *existence in the present moment,* their common possession of *matter's energy* in time, impacts the subject *cognitively* in ways that go beyond the subject's capacity as an abstractive knower, thus rendering the object intimately "familiar" in a way that knowledge could never provide. It reduces the distance between subject and object and makes the thing known to be "different" from an ordinary object of conceptual knowledge. The object is consciously embraced in its organic homogeneity to oneself — since both are equally constructed of *matter's energy.* This kind of reality is what I call the "really real." "Realization" is truly the proper term for it. Reality becomes "real" in this sense when we "realize" what it *means for us.* This "meaning" is *understanding.* The object becomes a partner in a relationship that cognitively impacts the subject more than mere "knowledge" because **the subject experiences its materiality, its real self, in the object.** It cognitively grasps a shared co-natural organicity that remains unconceptualized, therefore "un-known" even though more fully "understood." The object, in turn, engages the subject in a way that reveals the subject to be *related* to the object in ways that are both *prior to* and *more intimate than* knower-known. It is **the unity of the substrate,** *matter's energy,* that I believe is being apprehended in this experience. ... and while it resonates in consciousness, it was not the result of abstractive conceptualization.

In the example of the car trip across the continental United States, "three thousand miles" is the verbalized conceptual abstraction, the "known." The actual car trip, however, even though it will also be reported in the very same terms, "three thousand miles," includes the sights, sounds, food and lodging, human encounters, mechanical challenges, weather and climatic conditions, change in fauna and flora, landscape and roadway, fuel costs, elapsed time, effort, monotony and fatigue, that serve to place the subject in an intensified and *cognitively re-comprehended relationship* to "three thousand miles." It is the *experience* that makes it "real." Nor does this list have to be "known" in conceptual form or articulated as I have just done for the "three thousand miles" to be perceived differently upon being experienced. In fact, I had to articulate them as I did in order to even remotely *simulate* the actual experiences which, *on their own and without words and concepts*, would have put us across the threshold of *understanding* — made it "real." Thus it is the *experience* that adds a dimension to the concept "three thousand miles" without adding new concepts, or new information, or new data, i.e., new knowledge. *Experience* potentially places the subject in an entirely new *conscious relationship* to the original objects of conceptual knowledge.

understanding

The basis for all conscious *understanding*, human and non-human, is the cosmo-ontological *foundation*: the substrate, *matter's energy*. We are genetically, intimately, related to the world of *existence* because we are all made of *matter's energy*. We are immersed in what we are made of, and what we are made of is what we know, contemplate, interpret, realize, recognize and *understand*. The diaphanous connection between our consciousness and the world around us is itself a **sub-feature of the raw, physical homogeneity of existence,** *the universality of matter's energy*. We are all equally *present* and therefore equally present to one another in an immediate transparency because we are all made of the same "clay." We are the same "stuff," the same "thing."

We humans are not the only life-form to relate with consciousness and understanding to the world around us. This immediate organic interrelatedness works through a variety of conscious functions which we have in common with the animals. These functions do not depend for their characteristics on our peculiar human cognitive apparatus or on differences within the organic substrate. In our case this common ability is channeled into various pathways depending on how we as human beings have *decided to focus* our conscious awareness. Conceptualization, contemplation, interpretation, realization, and as we will see shortly, recognition and metaphorical predication, are not separate "faculties" indicating the presence of different "powers" or even of organic differences among human individuals or societies. They are examples of the *choices* we have made in the way we *direct our ability to consciously embrace* the reality in which we are immersed. The overall conscious embrace is what I call *"understanding."* It is a category of cognitive awareness characteristic of our species that is far broader than any of the operations into which it is subdivided, the primary one of which is knowledge. Our consciousness-as-*understanding* pervasively suffuses all our applications of intelligence. What I am driving at is that the operation we have defined as word-controlled "abstraction" is *only one way among many* others that we use to embrace our immersion-relationship with matter's energy, the *presence* of ourselves and others. And those "other" ways have a somatic dimension.

Verbalized abstract conceptualization, even though it represents, as I have suggested, the cutting edge of evolutionary development, organically producing the *specific* type of consciousness we call human, is still only *one way* that we focus our consciousness. But the general thrust of the capacity itself, *"understanding,"* we share in common with other life-forms, even though we exercise it in specifically different ways. Our primate ancestors and cousins enjoy a similar cognitive connection to reality. I am opting for a "vision" of humanity that is based *not* on the suppression of our abstractive rationality, but on its being put in its proper place, **as one useful function among many** to which we can apply our conscious powers. We relate to the reality around us with more than selective abstraction. The dethronement of word-based abstraction and the procedures of conceptual analysis that we call *logic*, does not mean their debasement much less their dismissal. But it does mean they lose their improper and overweening *exclusivity* as the defining elements of our *understanding.*[186] We approach reality in more ways than one. All of them lead to *understanding*, even if they don't all come from or lead to *knowledge*.

conclusions

These considerations reveal the breadth that human understanding embraces as it relates to the real world. To reduce *understanding* to "knowledge" defined as the abstract conceptualization of the potentially universal verbalized attributes of a given "object," is to minimize the question to the point of unrecognition. This, in my opinion, is the truly dangerous and destructive "reductionism" promoted by our spiritualist-become-rationalist tradition: it would *reduce* all value to the *immaterial,* which in our universe means restricting all value to certain human mental products and operations. For over two thousand years much of the philosophical and scientific enterprise of the West has been dedicated to the analysis of the abstractive features of consciousness for the purposes of transcendental projection (religion) or technical control (science). By first maintaining that abstraction established a transcendent immaterial relationship between transcendent immaterial realities (the "soul" with its intellect and the "essences" that reflected the perfections of "God"). It was then extrapolated to say that abstraction also established transcendent immaterial (cerebral) relationships to ordinary (material) reality, explaining and justifying technical control and exploitation of everything material, including that part of ourselves considered "material," i.e., *the body*. In our enthusiasm for that escapist illusion we have failed to understand ourselves, the organic survival apparatus with which we must function (our body-persons), and how intimately we are related to our material world.

---

[186] Heidegger, *What is Metaphysics,*p. 106

chapter 10
recognition

the meaning of meaning

In his book *Philosophical Investigations,* posthumously published in 1953, Ludwig Wittgenstein asks questions — lots of them — about our use of language. For some questions the answers seem so obvious that the questions appear vacuous. But, read with patience and taken all together, they begin to build to what I would call a philosophical quest for the *meaning of meaning.*[187]

You may notice in my use of that redundant expression that I must already "know" exactly what I'm asking about when I use the word "meaning" in order to even ask the question, *what is the meaning of meaning?* So why the question? What further "meaning" can meaning have that an answer to the question, "what do you *mean* by *meaning?*" will provide?

"Meaning" may be illustrated with a typical question of Herr W: "what do you *mean* when you use the word *is?*" If you make a statement like, "the rose is red," does "is" *mean "identical with"?* Of course not, he says. The rose is not the only "red" there is, nor is it normative. Could "is" then mean, "just happens to be" or "at this point in time" ...? That also seems to miss the mark ... so strict identity, like an equal sign, which is the usual definition of what the verb "to be" means, is not precisely applicable in either a permanent or a transitory sense. But we all know what "is" *means* in the sentence "the rose is red" even though we are at a loss to express it *in other terms.* In fact the attempt to explicate it *in other terms* is immensely more arduous, confusing and ultimately *produces less meaning* than just simply *using* the phrase "the rose is red" ... which gives us a clear and unambiguous *understanding.* We all know perfectly well what "the rose is red" *means.*

---

[187] Ludwig Wittgenstein, *Philosophical Investigations,* tr Anscombe et al, rev.4th ed. Hacker, 2009, Oxford, Wiley-Blackwell.

He then asks questions about the way we treat "*meaning.*" He asks, for instance, once we understand what a word like "is" in such a context *means*, do we remember what that "meaning" means, the way we remember, say, that the rose is red, i.e., the words? Isn't the *meaning of "is"* in that context, separate from any *picture* associated with the word? What is it, therefore, that we remember when we remember a *meaning*, and how exactly do we remember it without a picture?

words and pictures

Now words and mental pictures correlate. Our knowledge, we said, in chapter 5, begins with the sense image and aggregates like images through memory to form "concepts" — generalizations, universals — that stand for a multiplicity of like images. Our knowledge is a function of the recognition of likenesses within the context of material *existence* which is apprehended in "pictures." Without pictures, i.e., sense images, there can be no words. If the apprehension of *meaning* itself ... the recognition that makes it possible to aggregate like images ... has no image associated with it, it also initially lacks words.

It is only when we attend to the fact that we *understand* the connection between images, or the fact that we know a particular word-image *means a particular person* or thing or aspect of a thing that we create a new word — built metaphorically from "earlier" concrete words — that stand for these image-less recognitions. The very word "recognize" is one of them.

If "cognize" were a word, "recognize" would be a compound of it. It comes from *cognoscere* in Latin, which is itself a word built from *cum* and *noscere ... notus.* All these words mean to "*know*" (our basic English word, which itself is a transcription of *cognosere.*) We may never discover the original concrete image that lay at the base of the pyramid of metaphors. Is it the Latin word *notare* which means "to point to"? We can see that the words used do not explain nor describe the phenomenon; *they are only labels.*

I believe the reason that "recognize" is a compound of a compound is that the speakers who developed our language in ancient times, were attempting to *get beyond* any picture, any concrete image used as metaphor, and somehow touch this "immaterial" thing that we all understand intimately: the grasp of meaning. These linguistic acrobatics, despite their gangly awkwardness, still work for us because they are raw symbols with no *meaning of their own.* They work, ironically, because they are metaphors and we really don't need them. We all *understand* what it means *to understand,* to recognize. Words do not add anything to my *understanding.* But if I want to communicate, even with myself, (i.e., if I want to *remember* what I *meant*), they are a necessary evil. (I say "evil" because while, positively, words add nothing except an arbitrary conventional label, negatively they are the potential source of all confusion, misunderstanding and deception.)

Whatever the phenomenon *is* that the word "recognition" is trying to describe, it is fully understood on its own. No "word" can bring you any closer to understanding. The most that can result from *learning the word-label* is that this particular meaningless word, "recognition," is used to describe an experience to which you alone have access — your own act of *understanding.* You do not have access to mine, except as I use words that *evoke* your own experience.

recognition ... up close and personal

R*ecognition* is most commonly used to refer to our meaningful personal contact with another living individual. I say "living individual" rather than "human being" because the kind of recognition I speak of also seems to be functioning between some animals and humans and even between animals. In these cases "recognition" means that we are in contact with or at least aware of the *unique and irreplaceable "something" at the core of the living individual* ... what we refer to as their "individuality," or simply "personality." (Some call it a "soul").

"Recognition" is such a common phenomenon in the human world that it is hard to attend to its absolutely unique and inexplicable features. The experience of the "recognition" of unique individuals with which we are all intimately familiar goes far beyond the simple awareness of individual differences, no matter how unmatched and irreplaceable they may seem. "Recognition" seems to engage almost a different faculty of intelligence within us because it is the awareness that *the other*, like me, is itself a *power of recognition.* I recognize that "it" — the other — is recognizing me! What seems to be going on in this mutual recognition is that I am in such clear and intimate contact with *my own* ability to *understand* that I am instantaneously able to "recognize" that another also *understands,* and that both *what* and *the way* this particular person *understands* is different from the way *any other understands* ... **any** *other* ... *ever!* How is that possible? *Understanding* seems to be the same for us all. This, of course, refers to what is characteristic of our perception of our own personal individuality. Could it be that it is only my awareness of my own absolutely irreproducible uniqueness (can I even *really* be sure of that?) that I project of the other, **ever** ... and each of us *understands* such certitude to be a property of a unique personal recognition, even though, *rationally* speaking, there is no way that I could know in advance that *that particular person* ... that particular center of *understanding* ... or even *myself*, could *never, ever* be reproduced*!* In any case, all these recognitions — that of myself, that of the other as like myself, and yet unique, and that of the connection that binds the two and gives me *certitude* — is the phenomenon we are focused on in this short chapter.

There is something here that "goes beyond" the usual pictorial representations of concrete objects or events that we say were apprehended by "the senses," then formed into "images" by "the imagination," and finally retained in "memory" available for recall. Please notice that all these "operations" of human consciousness have been "assigned" to *different faculties* ... without there being any scientific evidence whatsoever that there actually are different organs or regions of the brain, or different chemical or hormonal combinations that account for these differing phenomena. Calling them *different faculties* is traditional, but there are no different "faculties."

Similarly, we "see" that there is an experienced *difference* between "recognition" and simple imagery formation and recall. On that basis, then, we erroneously assert that *that difference* means that "recognition" *transcends* materiality. That's an overreach. All we really have a right to say is that they are different. To claim that recognition goes beyond the capacities of *matter's energy* is to assume (a) that matter is inert and passive, (b) that something "beyond" *material energy* actually does exist and (c) that this is an example of it. We know none of these things.

I believe that what we are looking at is a repetition of the inveterate mistakes of our tradition. *As with the "faculties," we tend to imagine that different mental or emotional, or social events are the products of different and separate realities,* in this case a "soul." But there is nothing to prevent two apparently different phenomena from issuing from the same source. The human vocal apparatus is equally capable of speaking, singing, growling, screaming, laughing and crying. What is there that prevents "recognition" and image formation from being products of the same mental apparatus, as indeed they are.

Example: I "recognize" my friend Larry. My recognition is not separate from the sense image when I see him, or think of him in his absence. Now, one can try to say that the recognition is "*immaterial*" because it goes "beyond" the sense images, either present or remembered ... But then I ask, if it is "really separate" from the sense imagery associated with it, shouldn't I be able to have the raw idea, "Larry," without any imagery whatsoever? Try it. Try thinking of someone or something in the complete absence of any statement or memory or image of any kind. ... It is not possible*!*

I can have pictures of two people. One I recognize, the other I don't. What's the difference? What is going on in the "recognition" of the one I know that is not going on in the picture of the one I don't? And yet that "recognition" is equally had only in the seeing or the remembering of the picture or some other identifiable phenomenon going on in my imagination.

So if I insist my recognition ... which admittedly is "not just" a visual seeing ... is "different" from seeing, what evidence (even for myself) do I have for that ... or what apparatus other than my sorting through sense imagery and memories do I have for "*recognizing that recognition occurred*" in one case one but not in the other?

But there's more. I have a cat. And it's clear that my cat *recognizes me* when I go to the vet to pick her up ... What's going on in her "mind" that is different from what goes on in mine? And if these "recognitions" are different, how are they different and on what basis can I claim that such obviously similar displays are "really different"? I can't get into the cat's mind ... and even if I could, I bet that the only thing I would "see" would be *the image of me*, just like I see the image of my friend Larry in my own. The "recognition" is not "separate." It's a "dimension" or a quality that attends the imagery. It seems not to exist except in imagery.

This all recalls what Lonergan called the "non-sensate" quality of *insight*. *Insight* (recognition) seems only to occur in the presence of and focused on a concrete image ... but in any case it is not the apprehension of an image that is occurring, but rather its "recognition." The very same organic apparatus that is making the image is also instantaneously "recognizing" it ... but not always. Wittgenstein remarks, "... it is also possible for me to visualize a face, and even to draw it, without my knowing *whose* it is or where I have seen it." In this case, of course, there is no recognition. So when recognition occurs, what we "recognize" is not the image ... it's *Larry.* This is *understanding.* **It goes beyond conceptual "knowledge" but not beyond the capacities of material energy.**

Now, if "**knowledge**" is, as I have tried to distinguish it in this study, the system of word-labeled images warehoused in language for the purposes of group survival, then the phenomenon called "recognition" *goes beyond knowledge,* or may be said to be the *goal of knowledge.* The term I have chosen to represent the "recognition" that we are examining here, is *understanding.* Like the phenomena examined in chapters 7, 8, and 9 (interpretation, realization and co-existent contemplation), *recognition* is an *understanding* that goes beyond "knowledge." It is categorically distinguishable from the subject-object paradigm of human "knowing," and the very concrete images and their words used to evoke and communicate it. Recognition is certain, instantaneous, self-evident, self-explanatory and ultimately an entirely incommunicable accompaniment to concept (generic image) formation. It is the very *essence* of *understanding.* Because of the non-concrete / non-verbal nature of the phenomenon, its communication is **often freshly metaphorical** since it is a **unique event in the present moment of the connection of an organism with something other than itself.** Astonishing*!* That is recognition. It is one of the capabilities of *matter's energy.*

spirit?

Iisn't the description we have given of "recognition" the same as Lonergan's "insight," along with an implicit concession that this phenomenon is "non-sensate," and therefore a manifestation of "spirit"?

My response takes the form of a question: what gives anyone the right to claim that the non-sensate appearance of "recognition" is *not* simply a *property of* the apparatus we see functioning right before our eyes? What demands that something *other than* this organism, not in evidence in any other way, must be there to explain it? I believe this points up the prejudiced procedure whereby essentialists, and their cousins the reductionists, justify an ancient dualist world-view that they are committed to sustain *for other reasons* — the essentialists to insist on the existence of the individual "immortal soul," and the reductionists the inertness of matter.

As mentioned earlier, ancient Greek "scientific" thinking tended to *ontologize* (reify) heterogeneous phenomena, creating an *imagined "thing"* to explain their differences. It was routinely done with moral, psychological and social occurrences. Plato reified *ideas* ... Greek christians turned a Jewish workman-mystic named Jesus into a "god," and then erected his personal message and political assassination into a cosmic event. They concocted a cosmic "Original Sin" to explain human bodily urges, and they elaborated a ritual machinery (Church and sacraments) to "save" the world from the "devil" who became the reified substitute and explanation for the moral *anomie* and alienation caused by Roman imperial conquest and plunder. The "soul" I submit, is simply another one of these "ontological displacements." It is a non-existent "thing" conjured up by Greek "scientists" to explain certain unique human capabilities

We have no idea what the limits of matter's energy are, and therefore we have no justification whatsoever for claiming that certain phenomena, like "recognition" occurring exclusively in a material organism, transcend the capacities of matter, requiring the presence of "something else" to explain it. I contend that what we are looking at are simply the inherent powers of *matter's energy* released and activated by the particular organic configuration we call the human body, which was elaborated by evolution.

What exactly happens when matter "recognizes," we don't know. When we do find out, and I am sure someday we will, I'll suspect that that all we will "see" is just another very complex mechanism, one level down, that will still leave us wondering "why" and "how." It will have no explanation beyond itself. And the reason is, I contend, *there is nothing beyond it needed to explain it.* It does what it does because *that's what it is.*

I believe there is no more to know or discover. ***Matter's energy is a communitarian existential dynamism that has created everything by evolving it from itself.*** Consciousness in general, and specifically human consciousness as a development within that category, is *material energy's self-embrace* playing itself out in one of its — up to now — more complex and intensely unifying combinations. In "recognition" we can concur with Aristotle and say, *"the soul is somehow all things."* It puts on dramatic display the homogeneous "one thing" that is the substrate of this universe. It makes conscious unity possible, natural and even inevitable.

chapter 11
metaphor and *existence*

Despite what appear to be serious "criticisms" of word-labeled conceptual knowledge as detailed in chapters 5 and 6, it's not my intention to condemn it and the judgments based on it. I've already said that but I want to emphasize it. I am rather attempting to put that operation in its proper place. Effectively that means demoting "rational knowledge" from its up-to-now unquestioned ruling position in the assessment of human knowing — due in large part to the power of technology which utilizes it. The problem is not that concepts are invalid, but that their importance has been wildly overestimated. The scope allotted to *understanding* has been unduly limited to what I call "knowledge."

Exposing human knowing's earthly origins — its roots in the body and organic evolution — averts any illusory belief in its "divine" nature. I'm referring to the fact that linguistic conceptualization was once believed to be an operation only immaterial "spirit" could perform. We now know that consciousness is in fact an apparatus that evolved biologically for the purpose of survival — not unlike a weapon, or mechanism of defense. Human consciousness operates in the material world. It cannot be conceived in terms that affirm a false and illusory duality of mind separate from body.

But our efforts will be fruitless unless we simultaneously develop a new respect for those other cognitive instruments which allow us to pursue our conscious relationship to *being-here as material energy in the present moment* in ways that circumvent the distortions and limitations of word-based conceptualization. In the chapters immediately preceding this one, we examined four of the cognitive operations that bypass "knowledge," — contemplation, interpretation, realization and recognition. Here, I want to focus on *metaphorical attribution,* the principal verbal tool we have to defend ourselves against the stifling effect of social and linguistic convention in our attempt to apprehend and express our relationship to reality.

Not surprisingly this non-abstractive symbol-making activity, *metaphorization,* has been relegated to secondary status in the opinion of our technological and philosophical tradition. "It's only a metaphor," is a familiar phrase employed to dismiss a statement as less than objective, scientific and factual. The uses of metaphor will have to be rescued from the trash-heap of epistemic rejects and given their rightful place in the mental operations by which we relate to *existence.*

Metaphor is *linguistic innovation that is concrete and creative,*[188] and as such it offers the possibility of communicating more accurately the experience of *being-here physically* in the present moment, by which I mean *direct relational engagement.* It is the vehicle of *understanding,* i.e., meta-conceptual cognitive functions like realization, recognition, interpretation and contemplative reflection.

a. metaphor is creative.

Metaphor is *creative.* It eschews traditional word-labeled concepts and declines the use of conventional signification. The most powerful metaphors are fresh, unexpected, unusual. Much of their impact, in fact, seems to come precisely from a kind of shock, an *in*appropriateness, a dissonance that arises from the metaphoric descriptor standing at a conspicuous distance from the subject it is being used to describe. Take this line from the Song of Solomon:

> *"My love is terrible as an army set in battle array"*

We will recognize that the imagery in this metaphor is intense. It is also unexpected, and to all appearances, *totally inappropriate.* The gentle tenderness that is considered to characterize love is at first glance in jarring contrast with the terror inspired by the brutality of war, the work of organized armies. But on a second look, prodded by the author's suggestion of course, we may recognize the uncomfortable violence of love's emotions, the unintended, almost impersonal control that the beloved ex-

---

[188] In view of the current interest in metaphor and the many new meanings assigned to the term, I want to make clear that I am using "metaphor" in its general sense to mean a non-conceptual application of one concrete reality to stand for another. It includes the literary device called simile. So "Richard was a like a lion in battle," "God is light," "Possessions are an albatross." I call them all metaphor.

ercises over the emotions and decisions of the lover. We realize that love's power implies a loss of self possession that can produce a sense of helplessness. The metaphor evokes a melting readiness to capitulate as before a powerful army. Love isn't only joyful; it is also overwhelming, terrifying ... and possibly immobilizing.

Have I accurately reflected the intentions of the poet? At any rate my interpretation is what the metaphor, by engaging my own creativity, e-voked in me. I want to emphasize the metaphoric *process*: the freshly perceived existential significance of the metaphor-predicate, "army," to the subject, "love," impelled the poet to discard conventional descriptions as inadequate and, in the offing, to *challenge me to understand* his new perception. There may be more than one meaning behind the chosen metaphor. The poet might have meant for me to think about them all*!* The keynote is creativity on the part of both the speaker and the listener. There is a new perception which needs to find a new way to express itself, and demands an active participation of the audience *beyond the conventionalities of society's meanings encrusted in ordinary language.*

The new perception is transmitted in such a way that the metaphor generates a uniquely active and creative response in the individuals who hear it. There is no disembodied intellectuality, nor a peremptory, socially approved response functioning here. It involves a fully engaged communication relationship linking human being to human being, speaker to hearer, driven by the personal, individual revelation of the uniqueness of the present moment, that creates the metaphor. This relationship is at the heart of human communication and therefore, human community.

How does metaphor do this? The metaphoric image, in this case the "army set for battle," bypasses the conventional definitions that divide the predicate from the subject being described, "love." Metaphor is unexpected. It is foreign to the field of meanings generally associated with the word-object known. So metaphor forces us to engage our ability to perceive an object *as if for the first time.* It tends to precipitate a "realization" in a way that expands our understanding of the target entity beyond the conventional usage. After hearing Solomon's incisive metaphor, love should never be underestimated again. What we accomplish with the creativity of metaphor, is an intentional "return" to the original experience, *the present moment.* Metaphor consciously rejects the socially warehoused definition of things and charts a new course where every moment is new.

The reactions, then, imaged by metaphor, correspond to the integral perceptions of the whole person engaged in the uniqueness of the present moment, as opposed to the pre-fabricated and potentially detached reactions generated by conventional — socially forged and catalogued — words and their concepts. The kinds of perceptions that seem to require the creativity of metaphor for their expression are those that refer to what is *most real* for us, unique experiences that engage the relationship of the entire organism in the present moment. The resulting cognitive embrace is a "realization" — we *realize* what it means and so we *understand* it anew. The **somatic component**, often referred to as "emotion," or "feelings" and traditionally but falsely separated from "mind" and "thought," accompanies this appropriation and is fully integrated in the metaphoric take. The demonizing of the body-side of consciousness that accompanied the false belief that conceptualization is an immaterial, "spiritual" activity occurring in the mind apart from the body, is exorcized by metaphor. Thus it has aptly been used by the poets from time immemorial as a challenge to the imminent *existential suppression* lurking in the conventional usage of words. Metaphor, in other words, is chosen in order to preserve and communicate *existence accurately*.

human accuracy

It is customary to use the words "literal" or "factual," which imply "accurate," to characterize conceptual predication. That's unfortunate because it belies the reality. For it is exactly for the purpose of achieving *accuracy* that the poet uses metaphor. For the metaphor-maker, bypassing the conventional signification is essential to the precision with which the perception is both had and communicated. Embedded in this phenomenon, I believe, is an implicit declaration that the ordinary descriptors are not just simply less exciting or less artistic, but indeed potentially distorting, inaccurate, alienating. For the author, the use of metaphor is more than a decoration; it is a *necessary* tool of accuracy in these *matters that really matter*.

An army is an *army*; it has its conventional meaning. It's clear that this word and signification, before being given its metaphoric role, had nothing whatsoever to do with *love*. Once applied creatively by Solomon's poet to his "love," however, "army" is given a unique meaning, one never heard before. This "realization" returns both author and reader to that pristine creative moment, the awakening of specifically human consciousness, when words are first forged and focus human perception. From the author's point of view, a heretofore unknown existential significance of love *is actually being discovered and forged in the moment through the creative instrumentality of the metaphor.* Because of the primacy of *engaged relational existence* for human beings — the experience of the present moment — a special recognition must be given to the validity of the metaphoric application. It is essential to human accuracy.

Once the metaphor has been cast, we can say a "new word" has been created — an added signification for "army" and an expanded understanding of "love." Both words intermingle and are affected. If the metaphor happens to enter the verbal mainstream, the change may become permanent, and ironically the metaphor is transformed into conventionality. "God is Love," for example, has lost its metaphoric origin and is now taken as a conventional dogmatic concept. Metaphors which are used to represent human qualities, like "lion," "mule," "snake," have been so broadly accepted into ordinary speech that they have just become other words for brave, stubborn or sleazy.

All words had similar origins. The fact that a particular metaphor may never enter social traffic beyond the present moment does not subtract in the least from its validity. In fact, it is just the opposite: it is the very uniqueness of metaphor that gives it its power to return us to the present moment. Metaphor is a new creation, valid and uniquely expressive. In spite of the perennial assumptions, there is no justification for claiming that the conventional use of the word is more significant or accurate or valid than the metaphorical. This puts metaphor, I believe, where it belongs: on a par with other human instruments of expression in the search for truth.

But metaphor may be even more.

language and thought

A theory of the role of words in human ideation expands on this point of view. The hypothesis, known as Sapir-Whorf (SW), proposes that what we call "the ability to think," and the ability to use language, are two mutually dependent operations. If we define "thinking" as the ability to examine and analyze concepts, then no thinking goes on without language; language is an intrinsic component of concept manipulation. SW denies that we first form an "immaterial" concept or "idea" and then search for some material symbol like a word or a gesture to express it to others.[189] Language, according to this hypothesis, is *essential to the original formation of the concept itself*. The "idea" is a symbolic mind-body entity from the start and is generated only in immediate conjunction with its concrete material manifestation — its "word." It's an *"idea-word."* Words, in this view, not only permit us to express to others what we know, *they permit us to express it for ourselves.* It is claimed that without language we cannot know *what* we know or even *that* we know; without language there can be no reflexive self-consciousness. Language is what separates us from the animals.

While I agree that words have a determinative influence on thought, I believe the absolute nature of the claims made for them by this hypothesis goes too far. Concepts, in my experience, can also be wordless internal representations, *concrete images*, that aggregate singular perceptions. If words and concepts were one and the same thing, as SW proposes, it would be impossible to learn a second or third language. As we've seen, concept-formation is an ability we have in common with the animals, and animals have virtually no words at all.

But the SW theory, despite its shortcomings, serves to emphasize something that agrees with all linguistic observation: "metaphor" is at the very genesis of human thought and representation because it is a first and originating symbol created to stand for an internal idea. All concepts must have originally been represented by primitive concrete symbols — one concrete reality (a sound, a gesture, a line in the dirt) to stand for another — which is precisely our definition of metaphor. Metaphorization is the continued presence in us of the symbol-making power of human consciousness. The ability to create symbols, labels, signs, that externalize our internal representations of reality, made self-consciousness and therefore reflexive thought and complex human social interaction possible. Discursive reasoning and human society are equally the product of language.

A further theory assigns to human society the role of evolutionary selection for intelligence. The "Social Intelligence Hypothesis" (SIH) claims that it was society itself, by giving a survival advantage to those individual

---

[189] Stephen Pinker, *The Language Instinct*, 1994, Harper Collins, p.46 ff. Pinker describes it as "linguistic determinism, stating that peoples' thoughts are determined by the categories made available to them by their language." Pinker says, "it is wrong, all wrong."

organisms that *learned socially*, that "selected" for intelligence.[190]   Thus
the entire upward development of species with increasingly complex consciousness can be said to be a product of social choice, and language is
an essential component of societal development.[191]   Our earlier analysis of
the communitarian survival constructed by *material energy* since the very
beginning of time provides a corroborative cosmo-ontological background
for this phenomenon. Socialization mediated by language identifies a spi-
raling interactive dynamic that complexifies the human experience exponentially.

But society tends to "totalize," and create its own horizon. The storehouse of words which become the communal possession of society precede and therefore tend to pre-empt individual thought for future generations. Sapir-Whorf reflects this phenomenon. This explains the develop-
mental process by which conceptualization became more comprehensive,
more aggregative of earlier aggregates, more "abstract" (distant) in that
sense, and further from experience. "Thought" paralleled the developments in language as concrete meanings were expanded (and diluted)
into an infinite number of more subtle and socially processed derivations.

Let's look at an example: *Scribere,* a Latin word, means "to write;" it
has come into English only in its derived forms. One can imagine the long
line of developments among early speakers of Latin that must have taken
place from the initial use of the concrete root-word "to write" to its more
sophisticated and diverse applications: a-scribe, de-scribe, in-scribe, sub-
scribe, pre-scribe, pro-scribe, tran(s)-scribe and associated substantives
like script, conscription, description, inscription, prescription, etc. We must
also remember that writing was itself a multi-millennial work in progress
for the human species. From what primitive image did *scribo* itself originally derive? *Grabo,* to engrave? We may never know.

We can also see from this string of examples how metaphors become
incorporated into the mainstream of language by forming a figurative layer
of meanings, less concrete and more directed toward complex social or
psychological meanings, on top of the original primitive and more concrete
significations. I want to emphasize, however, that this figurative layer,

---

[190] see Appendix 9 on SIH
[191] Kerstin Dautenhahn *(2001) "The Narrative Intelligence Hypothesis: In Search of the Transactional Format of Narratives in Humans and Other Social Animals,"* Proceedings CT2001, The
Fourth International Conference on Cognitive Technology: INSTRUMENTS OF MIND (CT2001),
LNAI 2117, pp. 248-266, Springer Verlag.)

once incorporated as conventional meaning, is no longer "metaphor." For even though, for example, the word "describe" is a figurative extrapolation from *scribo,* and at some moment in the distant past had been the brilliant metaphoric invention of some creative proto-latin speaker, currently is part of the verbal storehouse of conventional meanings. We can discern its metaphorical origins, but it is no longer a metaphor.

So we can say that the metaphoric act is the living continuation of the creative process of *symbolization* — sign-making — that not only gave rise to our immensely rich language and social symbols but, according to some observers, reflexive thought itself. This makes metaphor more than a figure of speech; metaphor is the source of all human social interaction, the creator of human community. It is also the agent and measure of "truth." For what is truth but the correspondence between word and meaning — a correspondence *established by the wordsmith, the meta-phor-maker* enraptured by the present moment who creates a tangible symbol to express it.

For us, this is a new way of looking at things, because we have been trained by our technological culture to view *real existence* exclusively from the so-called *factual* point of view, rendered intelligible by the conventional concept and its word, and to consider other types of expression like meta-phor, as fable, fantasy, or artifice. But I am trying to suggest a correction: that human realities — and by that I mean the engaged relational realities that arise from *being-here materially, somatically, in the present moment* — are themselves a matter of unique and irreducible *fact* whose accurate perception and communication are appropriately grasped and expressed in the most original, creative, symbol-making process at our disposal, *me-taphorization.* And, the more evocative these expressions are of the rela-tions in question, the more real, the more accurate they are.

"feelings"

The prejudice against linking the word "accurate" with **relational en-gagement** is endemic in the West. We tend to give a priority to "mind" or rationality over what's been denigrated as mere sentiment, identified as "feelings," a reaction of the "body." To my mind this has been encouraged, if not precipitated by the Platonic paradigm which identified *the flesh* as the exclusive source of all such matters of "feeling." Feelings were *of the flesh*; they militated against rationality and hence they were dismissed or repressed. They were taken seriously only when it was believed they had a corrupting effect on morals, and then only to neutralize them..

By dropping *relational issues* from the roster of significant problems because of their bodily component, they were remanded to the limbo of the unexamined. The "accuracy" they required has been left to the determinations of popular culture. "Love," to take our current example, has been either analyzed abstractly (and incomprehensibly) by the ethical theoreticians in the schools and churches, or else it has been given meaning by the narrative metaphors of popular culture. Is it any wonder that we have two divergent definitions of "love," the one describing a responsible "act of the will" virtuously (and impossibly) devoid of emotional distractions, and the other an irresistible passion over which one has no control and little responsibility? Aren't such patent insanities as the Roman condemnation of contraception as *intrinsically evil* really an expression of the former?

Observe that the story-definitions of popular culture in fact predominate. They communicate because they are metaphor. They tell stories; they let one concrete reality stand for another. They reproduce engagement, experience. This corresponds to metaphor's second characteristic: *concreteness*, which we will examine shortly. Abstract predication does not communicate because, among other shortcomings, abstraction distorts concrete human reality.

This may seem all too obvious. But I believe we have not appreciated or explored the metaphysical implications of the power of metaphorical expression. Metaphor is not just a decorative device with entertaining imagery. I believe it is a standing clue to the primacy and ultimate human significance of *being-here in the present moment* in our Universe, the fully organic, bodily dimension of *existence* and the common substrate of *matter's energy*.

creativity, the radiance of existence

This analysis cuts across our presumptions about the exclusive *accuracy* of "scientific" or technological conceptualization as the extension of normal conventional predication and the apogee of human intelligence. If, as we are suggesting, neither conventional signification nor its scientific refinements can be expected to offer us a *completely accurate* assessment of *existence*, what we are saying is that we must come to rely upon the unstructured creativity, the innovative responsiveness of the fully engaged human being — the poet, the story-teller, the bard, the jester, the rebel, the lover, the mystic ... *in all of us* — to bring us closer to the truth of the *really real* world. Conversely, it also implies that we truly *count on* this creative power to help us see and express *in the newness of metaphor* what a given experience may contain, imply, and offer for *being-here*. Even though we have always valued poetry (which I mean in a broad sense of creative perception, expression and action), we must admit we "moderns" are unaccustomed to rely on such personal creativity, our own or others', for the direction of our lives. Poetry, like myth for the ancients, represents the ability both to see and to express familiar realities in unexpected ways. This kind of production cannot be programmed or mandated. We might even say it can hardly be expected, because it is the result of spontaneous creativity. That we *do* expect it, nevertheless, and that our expectations are not disappointed, is a testimony to the sustained creativity of the human species. Human communication is filled with the constant generation of new imagery and new language, sparked, in my view, by the frenzy for living common to all things, the self-embrace in which human perception encounters itself *being-here*, surviving, in the present moment. Creativity comes from sharing *matter's energy* and finding ourselves engaged, rapt, obsessed with *existence* beyond the power of convention to stifle or control. We know quite intimately that we "transcend" conventionality; we are related directly to *existence*. We are caught up in the throes of survival, the energy and passion for *being-here* in the present moment — what Spinoza called *conatus,* the drive to survive — the source of poetry and creativity and selfishness of every kind. This is authentic humanity and, in a sense that we are all intimately familiar with, it is not extraordinary at all. The life force is as fully in play for someone crossing a busy street as for Leonardo tantalizing us with Mona Lisa's inscrutable smile.

Recognizing the essential role of metaphor in our language does not denigrate the use of conventional meanings; it only suggests that we should keep them in perspective. The abstract word-labeled concept's main value is that it subjects the realities it classifies to predictability and social control by restricting them to *socially pre-digested, and accepted meanings*. There are important values functioning here. Conventional word-concepts are necessary for social cohesion and cooperation, not to mention the mechanical precision demanded by our vast and complicated technology. Those rational "truths," however, have been abstracted from reality and codified by society. Therefore, from the point of view of the individual *living in the present moment* they are partial and humanly inadequate. A major dimension is missing. By the identities being proposed here, since our abstract classifications do not include reference to the **existential fury** *of being-here* in the present moment, they falsify *actual* reality. The perspectives they respond to *are not fully human* and the results they yield are therefore equally partial. If these conventional meanings are utilized without the needed correctives that warn and protect us from their partiality, they will skew the human beings who think that they represent all of what reality is. They do not.

Such meanings have their "place." And we should keep them in their place.

b. metaphor is concrete

Metaphor is not only creative, innovative and unconventional, it is also concrete and non-conceptual. Metaphor does not "abstract." It eschews the use of *universals* as predicates and replaces them with concrete images — a real entity, quality or action is directly predicated of the subject: "My love is an army;" "time flies;" "Justice is blind." This serves to confirm our original critique of conceptual predication: it's an exercise in logic built on abstracted generalizations rather than an organically comprehensive and therefore "humanly accurate" picture of engaged reality. Conceptualization, we said in chapter 5, serves to classify those selected and abstracted aspects of reality that we, like the deer we looked at, have collectively learned help us to survive. But abstract concepts do not *comprehend* the reality in question, despite the traditional claims that they do. They rather *attenuate* it as required by current socially determined needs — like snap-shots — and then create an arbitrary symbol, a concept-label, a "word-image," to refer as shorthand to the snap-shots generalized as a group. Metaphor does not do that. By exclusively utilizing concrete images, metaphor lets a concrete entity or action stand, temporarily, improperly, inappropriately and therefore *provocatively* in place of a predicate, an appositional concept. "My love is like an *army*." It challenges the listener to *rethink* the issue. It upends the verbal apple cart by resisting the customary drift into conventional predication. It bypasses society's meanings and returns to the concrete image by using another unexpected concrete image as "predicate."

Returning to the concrete means returning to a pre-abstractive apprehension of the "object." An abstract thought may follow the rude metaphorical awakening. It is "rude" because it leaves reality undefined. In our example from Solomon, the target word, "love," is effectively "deconstructed," i.e., it is stripped of its conventional meaning and given a new meaning based on the new perception. It is returned to an undefined state, whole, entire, concrete and from there re-perceived. "My love," according to Solomon's poet, is not the subject of a virtue, or an emotion, or any other conventional definition of love; it is an "army in battle formation." Rude indeed ... what can such an outrageous comparison mean? Returning to the concrete necessarily engages the creativity of the hearers who are forced to "figure it out," to "realize" the significance of the brand-new dissonant connection established between the two objects by the innovator. It drives the hearers away from the conventional significance and forces them to form one of their own. Metaphor is thus always irreverent, always somewhat dismissive of larger society because it chooses to work outside society's language agreements.

Word-concepts are common property. The very purpose and function of concepts is that they are supposed to be agreed-upon meanings. The rejection of conventionality, therefore, as represented by metaphor, is an invitation to a personal stance. It calls for *thoughtful individual engagement* because the listeners must participate in the birth of the new interpretation apart from the considerations insisted on by larger society. We should observe that this engagement of the primordial creative power of language formation — which we've recognized to be the condition and source of human interaction and society — simultaneously *elicits the existence of new community*: gathered by the new meanings generated by the metaphors. Shared understanding creates "community."

The concreteness of metaphor conserves the sensory dimension ("feelings"), which characterizes all primary experience. Physical human contact with physical material reality is not suppressed nor transcended as it is claimed and believed to be in abstraction (a claim, by the way, that I challenge in chapter 5). Metaphor works unapologetically within the organic unity of human consciousness in its *immersion-relationship* to the material world, and preserves the integrity of the human organism from the false and illusory duality of mind separate from body.

Given the point of view being developed in this study, the *de*-conceptualization provided by metaphor is an absolutely necessary corrective to conventional predication. I accuse our tradition of having falsely interpreted verbalized conceptual knowledge to be the quintessential act of intelligence. In doing so it elevated to primary importance the "distance" involved in abstract conceptualization and made full relational engagement a secondary and inferior conscious contact. In effect, we glorified our survival apparatus, a saber-tooth, as it were, a raptor's claw, calling the human abstractive capacity immaterial, spiritual, and as a result **we divinized the most self-interested, distant, socially controlled and somatically alienated part of ourselves**. We attempted to ignore our bodies and were shocked to find that we were losing our souls in the bargain. Technological thinking has its place. But the unchallenged dominance of technological thinking as the exclusive vehicle for the accurate expression of truth is a distortion. It is a "cerebral" conceit that is both the source and the result of this distortion. It has meant, among other disastrous consequences (like the gross misinterpretation of what *existence* is *for*), that people are increasingly expected to function like the machines they create and maintain.

Existence is for *being-here*; it's not for going anywhere and it's not for doing anything, not even *seeing*. *Existence* is not about what humanity can achieve. **The achievement is already accomplished**: it *is* humanity. The recourse to metaphor is our spontaneous expression of what it means *to be-human in the present moment*. Being human is our way of *being-here-now*. **And being-here now, whether I'm doing anything or not, is what it's all about.**

chapter 12
altruism and society, ethics and politics

If we evolved as predators, as we surmised in chapter 6, how do we explain the existence of human society beyond the level of family and clan ... a phenomenon that clearly presupposes a general benevolence and cooperative spirit among people not related by kinship? How do we transcend our *instinctive selfishness*? This is more than an idle academic question, for it seeks the evolutionary origins of social interaction and the political structures that control our destiny. Evolutionary biologist Ernst Mayr says:

> One of the most important problems in the origin of human ethics is the growth of the hominid group from an extended family into a larger, more open society. For this enlargement to occur the **altruism** which before was reserved for close relatives had to be extended to non-relatives. [192]

Human beings, who supposedly are defined by a primitive individual "drive to survive," and then by an additional alienating adaptation to the life of a predator, nevertheless pursue community cooperation as a cherished value, and indeed complex society would be impossible without it. How is cooperation even possible with a "selfish" *conatus* focused on preserving itself ... and how could "altruism" evolve when natural selection favors the "winners"?

> If selection acts exclusively at the individual level, favouring some individual organisms over others, then it seems that altruism cannot evolve, for behaving altruistically is disadvantageous for the individual organism itself, by definition. [193]

---

[192] Ernst Mayr, *Toward a new Philosophy of Biology*, Cambridge, Harvard U.Press, 1988, p.78. Please note: the use of the word "altruism" was Darwin's. It refers to a biological phenomenon, not the psychological feeling and potential narcissism it often evokes. A better term might be "the cooperative instinct."

[193] Samir Okasha, "Biological Altruism", *The Stanford Encyclopedia of Philosophy (Winter 2009 Edition)*, Edward N. Zalta (ed.), URL=<http://plato.stanford.edu/archives/win2009/entries/altruism-biological/>.

This is a paradox that was not lost on Darwin himself, who argued in *The Descent of Man,* his 1871 sequel to *The Origin of Species,* that any **group** with altruistic individuals has a competitive edge over other groups without them, and therefore the prevalence of altruism is a function of *group selection,* not individual selection.[194] But there is a problem.

> The major weakness of group selection as an explanation of altruism, according to the consensus that emerged in the 1960s, was a problem that [Richard] Dawkins called 'subversion from within.' (*The Selfish Gene,* 1976). Even if altruism is advantageous at the group level, within any group altruists are liable to be exploited by selfish 'free-riders' who refrain from behaving altruistically. These free-riders will have an obvious fitness advantage: they benefit from the altruism of others, but do not incur any of the costs. ... By virtue of its relative fitness advantage within the group, the selfish mutant will out-reproduce the altruists, hence selfishness will eventually swamp altruism. Since the generation-time of individual organisms is likely to be much shorter than that of groups, the probability that a selfish mutant will arise and spread is very high, according to this line of argument. 'Subversion from within' is generally regarded as a major stumbling block for group-selection theories of the evolution of altruism.[195]

Whether altruism can be explained as a function of group or individual selection was a matter of considerable debate among evolutionary biologists from the 1960's through the end of the century, and reductionists like Ernst Mayr were persuaded that it could be actually explained by individual selection.[196] The debate may be of little real value because it does not consider the issue where I believe it really resides: *the substrate* that constitutes all things. Most of the reductionist arguments for the development of altruism include the prejudiced assumption that *altruism had to have begun with a* **genetic mutation** *because "nature" is presumed to be selfish*; they confine their work to the "remaining" issue: the spread of the trait throughout the population. I consider that assumption a *deus ex machina* produced "out of nothing" to justify reductionist prejudices. The explanation I offer here — the communitarian *conatus* — is thoroughly material and thoroughly explanatory. Since its intelligibility is grounded in the big bang itself, no mutations are required. I claim that *collective cooperation* is the very nature of *material energy — existence* itself — and that what has to be explained is, in fact, the reverse. The question is: since

---

[194] *ibid.*
[195] *Ibid.*
[196] Mayr, *op.cit.* pp 75-91 and 116-123. cf also the elaborate arguments of Hölldobler &Wilson, *The Superorganism,* NY W.W.Norton, 2009, chapter II.

individuals cannot survive outside of a community context, how and why do individualist traits get expressed, and how do we explain the increasing levels of individual autonomy as we go "up" the levels of emergence? Human individual autonomy parallels human intelligence. It grounds our universal demand for "freedom."

Before we get into this question, we should first be aware of the fact that altruism *is not just a human phenomenon. It exists throughout the animal kingdom* and so as a phenomenon it cannot be solely attributable to human social programming:

> ... Altruistic behaviour is common throughout the animal kingdom, particularly in species with complex social structures. ... some of the most interesting examples of biological altruism are found among creatures that are (presumably) not capable of conscious thought at all, e.g. insects. In social insect colonies (ants, wasps, bees and termites), sterile workers devote their whole lives to caring for the queen, constructing and protecting the nest, foraging for food, and tending the larvae. Such behaviour is maximally altruistic: sterile workers obviously do not leave any offspring of their own — so have personal fitness of zero — but their actions greatly assist the reproductive efforts of the queen.[197]

Studies of early childhood development as well as animal science and neuroscience all concur that "empathy," the basis of human cooperation and altruism, is a "hard-wired" feature of our organisms, built into genes inherited from our animal ancestors and functioning from the first moments after birth:

> ... babies seem to experience [empathy] in some form or another from almost the moment they are born. ... empathy does seem to be innate, and spontaneous acts of altruism on the part of babies are common (eighteen-month-olds will instinctively try to help a stranger in need though they haven't been taught to do so) ... [198]

But it is not only a human phenomenon:

> The empathy literature is completely human-centered, never mentioning animals, as if a capacity so visceral and pervasive and showing up so early in life could be anything other than biological.[199]

The same author, primatologist Frans de Waal goes on to explain that in 1992 the question of empathy

> "was resolved *at the cellular level* by the discovery of the existence of *mirror neurons.*" Primatologists "at the University of Parma reported that monkeys possess

[197] Okasha, *op cit.*
[198] Greenberg, *op.cit.* p.26.
[199] Frans de Waal, *The Age of Empathy,* NY Harmony, 2009, p.78.

special brain cells that fire not only when the monkey reaches for an object, but also when it sees another do so. ... What makes these neurons special," he says is that "they erase the line between the self and other and offer a first hint at how the brain helps the organism mirror the emotions and behavior of those around it. ... *The discovery of mirror neurons has been hailed as being of the same monumental importance to psychology as the discovery of DNA has been for biology.*"[200]

So the issue of altruism, initially broached in the context of human ethics and larger society, is seen to have a cosmo-ontological dimension because it *spans the disciplines* and different levels of organic life. In fact, its roots appear to be very old in terms of the evolutionary history of life on earth:

> ... empathy is part of a heritage as ancient as the mammalian line. Empathy engages brain areas that are more than a hundred million years old. That capacity arose long ago with motor mimicry and emotional contagion, after which evolution added layer after layer, until our ancestors not only felt what others felt, but understood what others might want or need. ...[201]

As with other examples of multi-emergence-level phenomena, altruism immediately suggests that an explanation lies in a "least common denominator" — some factor shared among the different levels of emergence. I suggest that common factor is *matter's energy* itself.

Animal altruism confirms that the phenomenon is far from being explained by our traditional philosophy which says it is a primordial inclination embedded in our *human* (exclusively "spiritual") consciousness in imitation of an altruistic "God." For the scientistic reductionists, on the other hand, animal altruism is especially perplexing because it runs counter to the (prejudiced) expectation that all directed biological energy will be an expression of exclusively individual (selfish) survival. The view promoted in this study is that the cooperative instinct is "natural" because it is a development of a primitive potential, a physical property of material energy itself. There exists in material energy *an existential communitarian dynamism* revealed during the collective formation of protons in the first micro-instants of the big bang, and repeated endless times and in myriads of ways thereafter. All structure in the universe, of whatever kind or magnitude, is ultimately the result of collectivization at one or more levels simultaneously. The *existential communitarian hypothesis* proposed and verified in chapter 3 explains both the *conatus* and its associated communitarian

---

[200] ibid., p.79 (*emphasis mine*).
[201] ibid., pp. 208, 232

character. And it does so with elegant simplicity, for it neither requires the outside intervention of a model "God," introducing imitative altruistic spiritual "essences" into "selfish" matter, nor does it involve the reductionist gymnastics of trying to explain altruism as a function of a still unexplained selfishness.[202] All development subsequent to the big-bang is a self-unfolding of the intrinsic physical properties of the dynamism we call *matter's energy* which in the context of an indifferent or even challenging environment *collectivizes* in order to survive. That it continues to do so throughout universal history and all its emergent levels, is no mystery.

ethics and politics

Acknowledging the material foundation of the communitarian phenomenon suggests a re-evaluation of our usual way of understanding the relationship between the individual and society and from there, ethics and politics. Politics is traditionally understood to be derived from ethics, just as society is from the individual. It was believed that the primary dynamic affecting human action in society and the resulting structures resided entirely in the individual.

The western traditions inspired by "The Book," grounded their ethical thinking in the relationship of the individual to an anthropomorphized "God" ... understood to be an *individual "person"* with a "will" who made human beings for a *purpose* and who requires that the individuals conform to that purpose by their freely chosen moral behavior. The ethics derived from the traditions of "the Book" is an individual program, directed as a personal command from an individual "God" to individual persons.[203] This individualism was later intensified by being wed to a Platonic definition of the individual as an *immortal soul* with an eternal *individual* destiny **apart from** all other individuals ... *and even apart from* his / her own body, in a world other than this one. Essentialism, as refined by Aristotle, subsequently made the individual, in the category of substance, the primary *locus* of metaphysical independence. Even the prescriptions deduced from

---

[202] The reductionists have a double paradox to deal with, since they also have to "explain" the drive to survive itself. They either simply assume that it was itself only a random variation or remain silent on the issue altogether. This is a major lacuna.

[203] I am aware that there was an ethical development recorded in the Hebrew scriptures whereby morality went from being an element of the collective relationship of the nation to its god to an individual morality as an element of an individual relationship. This is not the place to elaborate on that development much less to judge it or the almost universal agreement (in our individualist culture) that the development was an "advance."

"nature's purposes" called "natural law" remained defined as individual obligations.

Human community from this individualist point of view, is secondary. Societal structure draws its dynamism, cohesion and behavioral configuration from the God-bound, metaphysically independent **individuals** whose personal moral energy enlivens and directs it. *Ethics determines politics.*

priority of the community

The ethical implications of the cosmo-ontology elaborated in this study, however, are very different from the traditional, in content, in motivation, and in the relationship of the individual to the community. For we have made two very important discoveries: first, that the substrate of all things is "one homogeneous thing" and therefore makes the totality of things a unity. And second, almost predictable from that unity, an *existential* communitarianism dominates the survival strategy of *material energy* as it builds and launches its life-boats into the torrent of increasing entropy. Every emergent feature of the universe is the result of a cumulative collectivization. The **individual conatus** is *intrinsically and constitutively communitarian.*

The cooperative instinct, I submit, is simply the conscious, affective and behavioral resonance of the innate communitarian nature of the non-rational **conatus.** It derives from the physical structure of *matter's energy.* It is fundamental. It is intrinsic to our flesh itself. It precedes any form of consciousness.

Animal consciousness, in fact, by opening to the perception of the one and the many, produces the awareness of the subject-self and its opposition to others as object. It follows, then, that as consciousness evolves greater depth, scope and finally instrumentation (language), the sense of self becomes, to that same measure, greater. The higher the perceptiveness, the greater the degree of self-awareness.

This does not contradict the conclusion from part 1: that individuality, where there are no "essences," has no metaphysical basis and may be an illusion. Individuality must be recognized as *biologically emergent*, and therefore *functional with a subordinate status* (like a "vortex" within a rushing torrent). Even at the phenomenological level, the *experience of individuality* is a "learned" fact. It is posterior to organic *existence.* The individual is dependent upon *material energy's* earlier collective plateaus, including his / her organic base, the body, all of which have a prior and defining effect.[204]   Spinoza articulated this derivative relationship in metaphysical terms by saying individuals were the "modes" of one "substance." I use a different nomenclature, but the idea is fundamentally the same. The reality is the whole, and all other accumulated collectivities below it possess and exercise their own portion of this communitarian reality "downstream."

This way of looking at things inverts the traditional relationship between the individual and the community. There is no question of "forming community" in this conception, *for the community comes first.* The question is what kind of individuals the community forms, and how the community decides they will relate — to themselves, to one another, and to the collectivity itself — *politics determines ethics.*

This confirms our daily experience of life in society where human behavior and self-esteem, happiness, creativity, or conversely self-deprecation, immobilization and sadness, even suicide, correlate to one's perceived place within the community. The popular notion that imagines members to be rugged individuals, actually or potentially self-sufficient, independent moral agents who *contribute* their cognitive perspectives to a society which is simply the sum of those contributions, is pure illusion. As Durkheim and Kroeber insisted, the individuals are, rather, *a cognitive derivative* of society. *They reproduce society's ethical mandates and social goals.* They are driven by the community's motivations and constraints. And they are dependent on the sense of well being, *self-esteem,* that comes exclusively from social approbation — a *sine qua non* condition for individual productivity. Those who appear to function well outside of or in opposition to the parameters set by society, upon closer examination, will be found to draw their creative energy and self-esteem from a prior or parallel alternative society to which they are able to relate by direct association ... or in their imagination.

A logical corollary of this corrected understanding of the relationship between individual and society is that **ethics derives from politics** and not the other way around. "Politics" here does not mean primarily a power relationship, but rather a collective set of values, an *ethos.* Ethics is determined exclusively by the individual members' cognitive subordination to the needs of the collectivities to which they belong. The community decides what is right and wrong, exercising its responsibility to provide for

---

[204] Personality theorist Harry Stack Sullivan considered the individual personality a "hypothetical entity." Cf Hendrik Ruitenbeek, ed., *Varities of Personality Theory,* NY E.P.Dutton, 1964, p.122

the survival of each and all. This doesn't mean the individual disappears, an automaton in the social machinery. The individual exercises free moral choice and participates in the reciprocal shaping of the cognitive identity of the collectivity. But this shaping is always focused "upward," directed by the needs of the collectivities by whose existential survival power and therefore in whose service all "lower" collectivities exist. Individual ethics is subservient to human interrelationship in a kind of upside-down pyramid of levels: (1) family, then (2) "clan;" the village (church, clan, school) orients its members for service to (3) larger society, economic and political, and the larger society is subservient to its place in the (4) regional and global world ... which latter serves the (5) earth's network of life support systems ... all of it *in function of the survival of the totality upon which all other survival inescapably depends.* All morality is communitarian, and all morality is focused on the survival of the several collectivities to which the individual belongs. Morality as individual obedience to a "personal" God, can only be taken as a metaphor, a primitive poetry symbolizing the depth of feeling — the sense of "sacred" responsibility — that service to the community requires. But in fact, there is no "personal" God who demands obedience to "his" commands.

[*Author's note:* I am not dealing directly with the "reform of religion" in this study. But I would say that metaphor and poetry are an intrinsic part of any intense relationship. A fortiori, a relationship that is benevolent at levels that go beyond the personal, such as the relationship to the source of existence itself, will likely be described in terms of personal "love" and service. Such poetry is unavoidable, and maybe even desirable, given the intensity of the **conatus** — our love of life — and quite appropriate so long as it is always understood to be a metaphor. The "error" of religion, in my opinion, is to turn metaphors into literal concepts and claim that they have a quasi scientific reality. Whatever "God" there is, must be associated with the energy of the substrate in which all things "live and move and have their being." Apt metaphors used correctly can actually help promote that perspective.]

A sociological examination of the nature of society's constraints on individuals gives rise to the realization that *who we think we are* as moral agents depends almost entirely upon what the community we were born and raised in tells us we are. There is no abstract independent human being defined only by an "essence" of human nature, "intellect and will." The human individual is a painstakingly molded and highly detailed cultural and historical product, *formed and shaped* as to self-definition, moral and political vision, motivation, inhibitions and behavior by the communities and *community processes* where he / she lives. It is *education* in the broadest sense.

Sociology has known this forever. The all-inclusive formative power of the culture is axiomatic. The "wiggle-room" left to the individual after society's "nurture" has done its work is narrow indeed. Exceptions are extraordinary and necessarily "reactive," ... even **rebellion** is circumscribed by the social parameters in which the rebels were raised. This is another area, not unlike the material sciences and biology, where essentialism, locked as it is on the assumed spiritual characteristics of an "immaterial soul," does not explain, ignores, and may even deny the extent of the controls that come from social conditioning despite the avalanche of evidence for it. Reductionists, for their part, starting from inflexible premises that require that all behavior be explained as a function of a selfish *individual reflex,* do not even consider the possibility that the material substrate — to their mind dead, inert and passive "particles" — might actually be *intrinsically communitarian*.

Moral commands, despite all claims to the contrary, have always been created by the community to serve the needs of the community. Traditionally, the community elicited active moral engagement by *falsely* attributing commandments to a punitive and all-seeing personal "God," thus making individual moral agents out of their individual members. But it was a (self) deception. The commands are not "God's" and the obedience is not the individual's. *The community* is the true object (and subject) of the moral response.

work and survival

The cooperative instinct also implies *an egalitarianism* in matters relating to the economics of survival. Belief in an imaginary human "spirit" has been used since the time of the Greeks to justify *a stratification within society* (one of the so-called achievements of "civilization") in which the multitude (uneducated, "less-spiritual") have been forced to labor for food, clothing and shelter not only for themselves, but in order that the intellectual elite might be free for higher, "more spiritual" things. Hence the business of staying alive, *surviving,* which involves hard manual labor, has been considered a "bodily" activity, beneath the dignity of those dedicated to the life of the "spirit," and relegated to slaves. Within the memory of many of us manual labor was called "servile" and was the defining criterion for its prohibition of on "the sabbath."

Because "slave labor" is considered beneath human dignity the atti-tude has resulted in the abandonment of any serious effort to provide job-satisfaction in these areas, even by the workers themselves. People who have been taught to demean these tasks escape from them as soon as they can, contributing to a worse marginalization for those who, for one reason or another, have no choice but to take them. Metaphysical dualism may not have been exclusively responsible for creating class society, but it has performed yeoman's service in justifying and maintaining it.

To choose to escape from work, by our analysis, implies an alienation from *both the material and the communitarian dimension* of our core reali-ty. Such an alienation can only produce a deep personal frustration. We are *not* spirits! Our bodies evolved in an environment of the physical struggle to survive. Our spectacular achievements that seem so clearly beyond anything that "matter" can do, are in fact a dramatic display of the **spectacular capacities of matter** and how spectacularly blind we are to its transcendent creativity.

Our cosmo-ontology contradicts the anti-body fantasy *at its roots.* There are not "two-worlds," there is only one. Plato's philosophical theory of the two worlds provided a theoretical justification for exploitive stratifica-tion, the division of classes, and a sacralized individualism, and we are still burdened with it in our times. But the phenomenon doubtless existed long before the matter-spirit paradigm was conceived by the Greeks. Get-ting at the roots of it may involve rectifying *ideas* at levels even deeper than the Hellenic categories. [205]

work and equality

There is no need here to repeat a listing of the inequities in wealth and economic security that this exploitive stratification has meant throughout western history. I would simply add that by common experience, these inequities run directly counter to the communitarian instinct, having a ne-

---

[205] To describe the earlier journey that led from the communitarianism of our bio-chemical base and original hunter-gatherer societies, to the crippling individualism and class stratification of mod-ern "civilization" that produces so much personal and social pathology, we have to go through an extensive historical psycho-social analysis. For those interested in pursuing the precursors of the Greco-Roman Christian version, see Morris Berman, *Wandering God: a Study in Nomadic Spiri-tuality,* Albany, SUNY Press, 2000 especially the introduction and chapters 1 thru 4, pp 1-152. Berman sees the transition from nomadism to sedentary agriculture, producing an accumulation in excess of need that required "ownership" of land and goods and a method for their distribution, as the original source of both social stratification and ideological individualism.

gative effect on both classes. They generate a feeling of inferiority and resentment in the lower, who have traditionally been manual laborers and left poor; ... and a sense of defensive superiority in the upper, rendering life in society hopelessly divided, distastefully unconvivial, if not intolerably hostile. *Manual labor is a necessary part of human activity. The struggle to survive in a material world defines our existence.* Work is our destiny, the expression of what we are, not a "slave-class" burden to escape from.[206] Were it accepted as such, and remunerated at the level of its actual value to human fulfillment and survival, economic inequities would soon disappear.

It seems doubtful, however, at this point in our history that the mere appeal to change attitudes will have any effect on the inhuman conditions and addictive escapism that characterize manual labor after almost 10,000 years of "civilization." Any contrary vision will need massive "encouragement." A society that constructs itself on the foundations established in this study can easily contemplate including "survival work" in a *planned* collective program for the well being of all. This political "encouragement" would be a welcome first step in the rectification of the multimillennial distortion known as "class."

That work assignments might be *planned* by the community as a shared responsibility rather than coerced by the insecurities and escapist illusions of the "free market," is routinely condemned as *totalitarian.* That accusation speaks more to inveterate upper class prejudices and the false aspirations they arouse than to the repression of individual potential that a planned economy is accused of. Despite proclamations that "at least we're free," competition for "good jobs" produces no democracy for those employed in government, the military, churches, schools, as well in "free market" enterprises — businesses, hospitals, banks etc. Work in all mainstream organizations is assigned from the top down, and subordinates do what they are told. Claiming that competition for jobs creates a "free society" is a myth perpetrated by those who benefit from the insecurity engendered. The people whose full potential would not be realized in a planned society would be far fewer than those, impoverished and insecure, alienated from their own productivity and left to rot in addictions,

---

[206] cf Greek poet of the 6th century b.c.e., Hesiod, *Works and Days,* who speaks of the farmer who "... goes off to work as to a holiday." Also Karl Marx, following Hegel, declaring that work is a necessary (individualist) self-projection without which the human being remains undeveloped.

jails, on the streets or in foreign lands, by the free market tyrannies of the current one.

If today there is a growing realization that it is necessary to include every human being, and all species of life in the universal family, it's because we are *learning* that by serving the totality we are insuring our own survival. We are hoping that global ecological disaster can be averted because *we continue to learn* that we are an immense interdependent community. But the motivations often called "altruistic" are no more mystical or other-worldly than the quarks and gluons of which all things are constructed. The cooperative instinct utilizes the same drive to survive, and the same communitarian strategy that we first encountered in the dynamism of *matter's energy* coalescing into structured forms at the beginning of time.

western individualism

Human survival will be a communitarian achievement ... or no individuals will survive. Individualism, of which dogmatic *laissez-faire* capitalism is a derivative, is a false, misguided and self-destructive philosophy. It is the natural outgrowth of the dualism of the West in both its essentialist (greco-mediaeval) and modern reductionist (Cartesian) versions. [207] It is not natural. It requires the repression of the cooperative instinct. *Community is what allows us to survive*; we do violence to ourselves if we try to disregard that fact. The individualist repression of the cooperative instinct, I contend, explains many of our pathologies, individual and social. As Frans de Waal comments: "The evolutionary antiquity of this force [the cooperative instinct] makes it all the more surprising just how often it is being ignored." [208]

[*Author's note: such pathologies may have been the context of great creative endeavors like Nietzsche's bitter rejection of the christian culture in which he was raised; but our argument here is not that individualism destroys human creativity, only that it is unnatural and pathological. Creativity is an irrepressible property of the human being because of **the ability to imagine**. It is an organic potential that functions no matter what the cultural overlays. It is not the product of individualism, though it is often presented that way. The kind of creativity individualism inspires is most often antagonistic to it, and is invalidly adduced as an argument in its favor. We*

---

[207] The definition of individualism is varies widely among observers. I decided it would be best to allow the term to become progressively defined by the analysis presented here
[208] De Waal, *op.cit.*, p.232

*have no idea what creativity would produce in a communitarian culture, because we
cannot remember the last time we lived in a communitarian culture.*]

The physical, chemical and biological components of the human organism are communitarian. They create a semi-conscious pre-disposition to cooperation, but not a determinism. The human organism's ability to "imagine" what does not exist, has allowed it to shape those pre-dispositions, sometimes in self-destructive ways. *It is the imagination that makes us human*, and culture is a product of the imagination. This will help us understand how a communitarian substrate could have evolved an individualist social and economic system such as ours. *We imagined ourselves there.* The culture created by centuries of unchallenged acceptance of a repressive individualist ideology became Western "civilization."

In a version of "individualism" that we are familiar with today, our *culture* teaches that happiness is an individual achievement and possession, and that society's role is to protect the individual's ability and opportunity to achieve happiness. In more extreme versions it is considered perfectly acceptable and perhaps even desirable, that the competition that allows the "winners" to achieve, means the actual physical extinction or economic destitution of the "losers." Such attitudes are consistent with the existence of a menial underclass, marginated and miserable.

How did this come about? Allow me to suggest an hypothesis: I believe that such a conviction is the ultimate product of a cluster of ancient beliefs that *were centered on the existence of* an individual "spirit-God," who rewards and punishes each individual "spirit-person" separately in a "spirit-world" after death. In this view there was no group survival. Everyone was on their own. The intense concern for oneself and the disdain for one's material body in this material world that this vision produced was introjected deep into the western psyche. It endured long after the religion promoting it stopped being the principal object of belief.

Rome

Specifically, my contention is that it was **christian religious ideology,** already an amalgam of Jewish narrative and Platonic Greek philosophy, wed to a **theocratic Roman Empire** 1600 years ago, that created western individualism. This hypothesis may seem counter-intuitive because we are accustomed to think of "religion" as the source of community. "Love one another" is a central commandment. But the communitarian "commandments of Jesus" (which were in reality the restatement of the Mosaic

Code), effectively became a voluntaristic, if not romantic overlay, super-imposed on a Greek *individualist* philosophical foundation functioning in the service *of Roman Law* with its emphasis on reward or punishment. Looking at this phenomenon more closely will help us understand how *individualism* became the central dynamic in the European lands ruled by Rome.

The tolerant become intolerant

It is significant that, prior to the appearance of the christian phenome-non, the Roman State was notoriously tolerant of all kinds of religions. In fact, foreign gods were embraced enthusiastically and incorporated into public worship along with the traditional Roman deities. It was an expres-sion of the empire's universalism and one of the ways Rome earned the good will of its conquered subjects. Two examples of this open-armed acceptance were the cults of Isis from Egypt and Mithra from Mesopota-mia, very popular in Rome and with the soldiers of the legions in the first century c.e. There was never a persecution of any religion, including the rebellious Jews, *until the Christians appeared.* Rome persecuted Christi-anity within decades of its appearance in the Greek-speaking world. Why?

The Roman authorities did not attack Christians on religious grounds; Christians were persecuted because they were seen as a threat to impe-rial power. By refusing to sacrifice to the gods, they challenged the legiti-macy of the theocratic state. The third century persecutions of Decius (251 c.e.) and Diocletian (302-313 c.e.) were explicitly political and di-rected at the *subversion* inherent in the christian refusal to acknowledge the divine status of the empire and therefore Rome's divine right to rule. The martyrs who did not sacrifice were thus effectively (though not dog-matically) *anti-imperialist because* they professed their loyalty to a differ-ent community — a different center of power.

The christian attitude toward the empire changed dramatically after the Constantinian ascendancy. Once Christianity was officially sanctioned, the empire became the "different" community. *Traditores*[209] were treated with great leniency by the newly elevated "Catholic" authorities. Groups like the

---

[209] **traditores** were those clergy who "handed over" the liturgical books and scriptures to the Ro-man authorities in return for personal safety. They had to be higher clergy, for they were the only ones who controlled the "books." It was a specific demand of the Diocletian persecution. Our Eng-lish word "traitor" comes from there.

Donatists in North Africa who refused to accept the reinstatement of *epis-copi traditores* were declared "heretics" by the "Roman" church and violently suppressed by the legions of the Empire.[210]

Once Christianity was installed as the State Religion, it became a tool for the power projection of Rome in a way that the older pagan cults had not. Hence, after Constantine, religions other than the official "Catholic" version were systematically exterminated. This was unheard of earlier. The Jews in particular, claimants to the authentic "ownership" of the Scriptures, became the target of attacks that had a genocidal savagery. Rome had never before persecuted the Jews. But doctrine had become an instrument of conquest and control. And one of the ways Christian doctrine worked for Rome was to neutralize any community (= power center) that was not nested within the Empire's Church with its individualist ideology of guilt, sin and eternal destiny in another — immaterial — world.

Community is survival, and so community is power, because survival is non-negotiable. It was in the empire's interest to eliminate all centers of power except its own. Roman power was consolidated with unparalleled efficiency by a Romanized Christian religion that made each person an isolate, terrified for his / her own eternal destiny in a world other than this one, and dependent on the Empire's Church for salvation. The Empire did not *create* an individualist ideology in the Christianity of late antiquity; it was an autonomous development of Greek thought that Rome exploited and expanded.[211] Once Christianity became wedded to the Imperial State its vision ruled the "whole world," *kat'òlon* ... hence the imperial community was *kat'òlica.*

*[**Author's note:** This analysis is not offered as religious criticism or reform. Its intention is **to understand** the impact of Christian doctrine on the development of western culture leading to the formation of our particular individualist view of the world. Whether the political effects of Christian doctrine were consciously intended, or whether the teachings in question represent a corrupted version of Christianity, or that "true Christianity" was being practiced by a "remnant minority" in monasteries,*

---

[210] Constantine violently suppressed the Donatists in 317. Their continued existence prompted Augustine to ask the Emperor Honorius to send the legions again in 405.

[211] A thorough discussion of the evolution of christian doctrine from the communitarian vision of the first two centuries, to the individualist form it assumed in late antiquity, is beyond the objectives of this study. It is enough to say that the historical factor driving that evolution was the increasing hellenization of christian theology under the influence of Platonic and neo-Platonic categories, specifically the "immortal soul" which was *not* part of christian doctrine in its original form. Cf *An Unknown God* chapter II.

*or by officially marginated "heretics," is not my concern here. I am interested only in the influence on western culture: "who we think we are" and what we think **existence** is, as shaped by these doctrines and their political utilization.*

*Along these same lines, how exactly Rome **re-shaped** Christian doctrine to make it an apt instrument for Imperial needs is not the topic here. But in this regard I would cite one striking example: Constantine's personal involvement in the dispute over the divinity of Christ. Keeping in mind that he was not a Christian, the Roman emperor's investment was on conspicuous display in (1) his personal convocation of the first "ecumenical" Council in 325; (2) his choice of venue at his own villa in Nicaea; (3) his attendance and detailed participation in the deliberations of the Bishops and (4) his personal insistence on the use of the word **homoousios** to define the relationship of the Logos to the Father in the Trinity. Constantine's "word" was reluctantly accepted by the Council Fathers despite their universal distaste for it as unscriptural. I leave it to your political imagination to discern the reason why the Emperor of Rome would be so intensely involved in resolving a theological dispute ... and how Christian Bishops could be capable of permitting not only his participation, but of acceding to his preferred solution, which was not theirs. Cf A.H.M. Jones, **Constantine and the Conversion of Europe,** NY Collier Books, 1962 reprint of the original 1948 text published by MacMillan. It has been republished by The U. of Toronto Pr numerous times from 1978 to 2001.*

*To get a sense of the theocratic mentality of the Roman world and the assumed identity of the State and Religion cf also Eusebius Pamphilius of Caesarea, the famous author of **Ecclesiastical History** who also wrote a **Life of Constantine,** (VC) in 339 c.e., the year of Constantine's death. He considered Constantine a saint.*

Eusebius was one of the Council Fathers at Nicaea and describes the deliberations in terms that evoke an image of Constantine arbitrating a dispute between litigants ... hearing both sides and suggesting a resolution (VC III 13). Constantine's presence and the triumphal closing of the Council, timed to coincide with the 20th anniversary of his reign, was obviously an awesome event for Eusebius. He ended his account of the Council by describing the banquet which all the bishops attended as

> *" ... splendid beyond description. Detachments of the body-guard and other troops surrounded the entrance of the palace with drawn swords, and through the midst of these, the men of God proceeded without fear into the innermost of the imperial apartments, in which some were the emperor's own companions at table, while others reclined on couches arranged on either side. One might have thought that a picture of Christ's kingdom was thus shadowed forth, and a dream rather than reality." VC III 16*[212]

imperial theology

"Original Sin" is the foundational "christian" doctrine establishing *individual selfishness* as endemic to, if not the very definition of, the human

---

[212] Eusebius Pamphilus of Caesarea, *The Life of The Blessed Emperor Constantine*, tr. Bagster repr. Grand Rapids MI: Wm. B. Eerdmans, 1955.

being. It is significant that this teaching was elaborated in its most toxic form by a member of the Roman elite of late antiquity, Augustine of Hippo, a century after Constantine's "donation." But note: It was a full 350 years after the birth of Christianity, and could hardly lay claim to early authenticity.[213]

"Original Sin" institutionalized individualism on three levels simultaneously. (1) It taught that every individual human being was corrupted by Adam's disobedience; and it declared that each individual was held **personally** guilty for the affront to "God." (Even newborn babies, if they died without baptism, would be damned for all eternity.) (2) Personal corruption entailed a proclivity to sin, which meant an innate **selfishness** — an inability to consider anyone but oneself. And (3) sin meant **individual damnation**, unless sin were forgiven and remedied — individual by individual — with the ritual ablutions administered exclusively by the Empire's Church. It was only with Augustine's Roman imperial theology that we hear "*outside the church there is no salvation.*" It gave the empire all the justification it needed for external conquest and internal control.[214]

In tandem with the *individual immortal soul,* "Original Sin" was perfect for their purposes. Your "selfish body" would always make you sin, and you would always have to go to them for forgiveness ... terrified of dying and going to hell, you knew that there was no one **except them** who could help you avoid eternal damnation. They made the primary human survival communities in this world irrelevant; and they put the Empire's Church, offering survival in another world, in its place. To effect that transition, they had to convince you that you were a spirit from another world, that you belonged to no one but yourself, and "God" in that other world ... *and they owned "God."* Family, clan, village and region lost their significance for material survival, overshadowed by the "spiritual" survival threat coming from the "other world," and were absorbed into the community of the

---

[213] The early source cited for "Original Sin" is the NT Epistle to the Romans. But what Paul expressed there did not in any way justify where Augustine went with it. Augustine's version is thoroughly hellenic, individualistic, a reprise of the neo-Platonic theory of the fall of spirit into flesh, denigrating matter and permanently institutionalizing individual human guilt. Paul's was hebraic, communitarian, symbolic, using the story of Adam's sin to evoke the re-creation of humankind accomplished by Christ alone — establishing a new relationship to "God" **for all people,** *without law, guilt or sin.* Augustine completely reversed the meaning of Paul's imagery.

[214] 16th century Christian colonization of primitive peoples entailing their exploitation and enslavement was universally justified as "christianization." Imperial Christian Doctrine has always motivated and justfied conquest while rendering the conquered submissive — a lethal combination.

theocratic state, once declared dead by the martyrs and now brought back to life by the kiss of the Imperial-consort. And all of it was based on the central dogma of the spiritual individual, the existence of the "other world," and the corrupt nature of everything in this one made of matter.

Imperial Roman "christian" theology established the premise that material nature is inherently corrupt and selfish. It would follow, then, that any attempt at forming "human community" outside of the salvific community of the Empire's Church must be the product of some kind of *selfish materialist dynamism* ... if it claimed to be altruism, it must be *narcissistic,* for without the Church all is material, all is corrupt, all is selfishness. Doctrinaire capitalism, as the ultimate violent self-projection of the individual who arrogates to himself as "owner" what belongs to all, is inevitable in such a world. For the human being is intrinsically selfish and insatiably greedy. *That cannot be changed.* The best that can be done is harness and direct that demonic energy.[215] Besides, those who are convinced that they are utterly despicable before God, develop a need for a compensatory mechanism that will allow them to "prove" their worthiness. The accumulation of wealth as a sign of divine favor was a solution that came to be institutionalized by the Calvinist version of Christianity in the 16th century. We still speak of the well-off as being "***blessed by God***." This generated a belief that the wealthy were *morally superior — elected by God to teach and to rule.* The plutocracy that today rules the world came from there.

The insuperable individualism that we have been brought up to believe is the norm, is the result of a very long, intense and continent-wide cultural programming in the lands we know as "Europe." For more than a thousand years, Europeans were subjected to this Christian indoctrination telling them who they should think they are and what they should think of this world we live in. The result was to breed a continent of isolated individuals, vulnerable, defenseless, terrified of "God," distrustful of themselves, their neighbors and this material world along with their own material bodies. Even if there were some who sensed the inhumanity of it all, they had no alternatives. The few options that did exist, like the old paganism, Judaism, Catharism or other heresies, were the targets of extermination by the state.

---

[215] Cf Thomas Hobbes, *Leviathan,* 1651: "man is a wolf to man" "... I put for the general inclination of all mankind, a perpetual and restless desire of power after power, that ceaseth only in death." "... In the state of nature profit is the measure of right."

The doctrines of the Empire's Church formed the soul of Europe and its political structures. The Western mind was thoroughly conditioned along lines that meshed with Imperial requirements. Western colonial conquest over the past four centuries, energized by this vision, has exported that mindset and insured its globalization. At this point in time, the formality of modern religious "emancipation" has done virtually nothing to undo the ingrained personal definitions established with such unassailable authority and deep penetration by these doctrines over such a long time. European capitalism and its associated pseudo-democratic plutocracy is the product of an ancient imperialist ideology; and in our times they have come to dominate the planet.[216]

in contrast ...

The communitarian instinct is a physical property of *matter's energy*. It is only secondarily a moral commandment and an emotional preference. The primordial "law" of collective survival is violated when people stop working rationally together for the survival of each and all. Our culture told us that we were *individuals* and that each had an eternal personal destiny separate from everyone and everything else including our own body and the organic matrix that sustains us. Individualism denies that we belong to one another and to the earth.

Some will object: if altruism is such an innate instinct, how is it that a global economy like ours, built on individual self-interest, has emerged and is thriving?

Thriving? Deaths from armed violence, disease, crippling poverty, starvation and displacement, all born of massive economic inequities within the human family are the lot of the *vast majority of humankind* ... and every bit of it preventable! Add the extinction of species and damage to the earth's life-support capability from an exploitive technology unconsciously driven by the same defunct, dysfunctional individualist ideology, and I submit that this system, built on self-interest, is in fact an unqualified disaster.

This study proposes a philosophy based on the discoveries of science that attempts to elucidate the nature of *existence*. It is not trying to sell a

---

[216] Empires existed before and beyond the reach of Rome. The psycho-social dynamics of their creation and evolution would have to be studied separately. In our times, however, after 4 centuries of western colonialism, the form of all non-western power projection has been strongly influenced by the European mindset. It is validly considered global.

social program, even though the broad outlines of one is implied. There is no *message* except what flows from reality itself. If this proposed philosophy is correct, it would seem that basing social choices on individualism would generate a deep-seated frustration leading to personal dysfunction.

Moreover, the dualism that underpins individualism, by denigrating "matter" as dead and inert, encourages attitudes and activities that may end in the destruction of the environment. This is not a sociological analysis or a political pamphlet. It is a philosophical enquiry into the nature of reality, and it is saying that the communitarian nature of reality *explains* the frustrations and dysfunctionality that we have all experienced ... and it predicts that if we don't change course, the networks that support us — our society and our planet — may collapse entirely.

I repeat. This study is philosophical. Like science, it seeks to understand reality *such as it is*. I am not exhorting readers to *imitate* a communitarianism discerned in the construction of things throughout universal history. I am saying we human beings *are* one of those things and therefore we have an *innate communitarian instinct,* intrinsic to our very *physical existence*. It is our **conatus.** I am not saying "let's make an effort" to be altruistic. I am saying the cooperative instinct is embedded in our bones ... we may have noticed *how hard it is to suppress it.* It's what we are, we can't get rid of it. We should stop fighting it in the name of an individualist fantasy-world that does not exist.

part 3:
understanding existence

chapter 13
the limits of knowledge

being-here is energy

W e're trying to elaborate a replacement for metaphysics, which we are calling *cosmo-ontology*. In this chapter we will review what we have learned about *existence* and see where it may lead us. Our goal is to *understand* reality. A reminder: we are using *presence, being-here,* to refer to the experience of *existence* in the present moment. *Existence-in-time* replaces the term "being" which we have accused of being a false conceptual construct that skews our perception of the true characteristics of reality.

We have also determined that whatever *exists*, as far as we can know, is *matter's energy*. Ultimately it is reducible to *energy* which is the primordial reality. I use the doublet *matter-energy* to avoid any temptation to separate the two. They are one and the same thing. Moreover, we have said that *matter's energy* **is existence**. Material energy is what we are made of and it's what everything that we can relate to is made of. We can have no direct cognitive relationship with anything that is *not* matter's energy. Even mediaeval philosophy, dedicated to establishing and exploring the world of immortal spirit, admits that knowledge of such transcendent realities is *necessarily indirect,* an inference made from *the only direct natural knowledge we can have*: the experience of material things.

The foundational ground for the possibility of our awareness of everything, both ourselves and things other than ourselves, seems to be that we are all made of exactly the same "clay" — constructed of the same sub-atomic quarks and gluons, electrons and neutrinos which some believe to be simply the different *vibrations* of homogeneous strands of energy called "strings." What appears to us as inert solid matter, is simply a different manifestation of this energy.

Existence is energy; that means *being-here* is not simply *here*. Being-*here* is not at rest; it is intrinsically **dynamic.** It moves, it changes, it enters into combinations within itself which modify its activities and its appearances. It *selects* among the features and character created by these new collectivities on a continual basis — always in the service of only one goal: **continued existence**, survival — survival means *existence*. This restless recombination defines material energy's evolution — always changing, always in motion, always in process for more existence.

Existence is *matter's energy*.

The process that emerges from material energy's dynamism uses repeated patterns of recombination that we have called a "communitarian strategy." It is focused always on *being-here*. We say it is an "undefined" energy not because it has no direction, nor because it is formless, but because the form it takes is not heuristic, i.e., it is not regulatory or guiding. Form *follows* this energy; it does not lead it or direct it. This constitutes the seminal difference between the modern and ancient perspectives on metaphysics. **Being here is only directed by and for being-here.** *It has no purpose beyond itself.* It is absolutely self-determining and all forms are subservient to *existence*. Rational thought, *plans, purposes* do not describe or define this process. **There is no point to being-here except to be-here.**

We experience *matter's energy* as an *existence* that is driven to endure. This goal remains ever the same whatever the recombination. All its many changes are for only one thing: *being-here-now*. *Being-here*, therefore, means *staying-here*, continuance, and so it implies perdurance in time. The attempt to perdure spawns a necessary derivative: *presence* resists cessation and dissolution. It survives. So perdurance necessarily implies *being-here* "better," that is to say, more securely, more tenaciously, more intensely. It is the foundation of *survival*. It explains the *changes* that produce new species when environments change, and it also explains the extended *stasis* (resistance to change) characteristic of successful species when environments *don't* change or don't change enough to warrant adaptation. Being-*here* is *a passion for itself*, an obsession and an insatiable addiction. We have called it a **congenital self-embrace.** *Presence* wants to endure, but not simply to continue; such craving seeks a guarantee and therefore an intensification of what it does. It wants *to be-here*; it wants to insure *being-here*. So it is driven to survive, to embrace itself in a paroxysm of self-possession. Consciousness is only one of the many manifestations of this self-possession.

Presence is-here and what is *does* is to stay here. It *survives.* It is what Spinoza called the **conatus sese conservandi** — *the drive for self-preservation.* In Spinoza's vision it was the core property of everything that was ... a modal expression of "God" from which everything emanated — a "God" whose "essence" was *existence itself **... esse in se subsistens**.

So *matter's energy* which expresses itself in *being-here* displays itself as a self-embrace, a thirst for *being-here* that goes on and on in time and in intensity. The metaphoric nature of the description offered here is intrinsic to our interpretation. We will deal with the significance of this shortly. But here it's important to emphasize: the drive manifest in this perdurance is not the result of evolutionary selection. Selection, rather, presupposes it as the source and explanation of its effects. Natural selection is an expression of the *conatus.* It is the basis of all development and therefore we can also say, *it is a function of matter's energy.*

existence is time

The notion to which we have given the word-label *existence* is not derived from an abstraction. It comes from the experience of *being-here-now.* If there is any valid "intuition of being," it is here and now that we find it. The experience is that of the *present moment* because nothing that exists, exists in the past or in the future; whatever exists, exists only now. It's a "now" for which the essentialists with their obsession for eternal immutability have little respect. For the "now" we experience is not a fixed value; it is a fluid, changing, temporary phenomenon; it is always gliding out of the past and into the future. *Being-here* is essentially time-related; it is a *modulating process.* For since *being-here* manifests itself in the present moment, the perdurance of any entity comprised of *matter's energy* necessarily creates a flow of present moments, a non-discrete continuous sequence proceeding into the future. The insistence of what's *present* to remain *present,* which is its self-embrace, creates our experience of time.

This flow of time is inaccurately said to be composed of "moments." Reality is, in fact, an unbroken *continuum* perceived by our minds as static entities enduring through the sequences of time imagined as "moments." We use that term "moments" only because we find it difficult to imagine pure ceaseless unpartitioned change. Our concepts, we are reminded again by this, are like snap-shots. They freeze selected aspects of incoming data. It's the way abstraction works. We cannot immediately "conceptualize" *time-as-endless-flux* even though we experience it that way. Fortunately, we are able to refine our images because we reflect on experience and so we can intentionally work the fluidity of time into our notions. But there really are no *instants* or moments of time.

There are still other corrections to be made. Our images don't always conform to the phenomena. The perennial philosophy tended to imagine *existence* as if it were something in itself apart from what exists. (That's because our word-labeled "snap-shot" concepts — our ideas — tend to be taken as if they were "things" and not mental images. It is the basis of Plato's fatal error.) Our ancestors also erroneously conceived spacetime as if it were something independent of the enduring existence of what *is-here.* But time is not a glitch on a graph, or the tick of a clock. Spacetime does not exist apart from what survives and endures, nor does *presence* exist apart from the particular configuration of *matter's energy* actually surviving as this or that individual entity. Time and temporary configuration are simply the way we experience energy gathered, *being present* and remaining *present.* Time is the perceptible continuum produced by what *being-here* does. *Being-here* embraces itself and its integrated functions, its recombinations. It endures, it transcends the moment and carries itself endlessly into the next — it *survives* as itself. *Time* is simply another word for the experience of *being-here-in-process*, existence sustaining itself, clinging to itself, and changing as it must, to remain itself.

This understanding of time as a derived property of *matter's energy,* as we saw in chapter 3, corresponds to a similar understanding of *space.* It concurs with the new understanding that has emerged from the theories of relativity about the unified phenomenon now called spacetime. Spacetime is not an "entity" in itself. It is the measurable perception of the relationship of *matter's energy* within itself, to itself, as an *existential self-embrace*; it is a derived property of *the **conatus,*** the inherent self-sustaining dynamism of the substrate. The notion that *material energy* even in the form perceived as *particles,* can be understood as a *field* of presence that extends, not unlike gravity, throughout the entire universe.

This further emphasizes the unity of *matter's energy* as a Totality. What we see when we look out on the Universe are not discrete, independently existing particles or their aggregates residing in an empty "clockbox" called spacetime. We are looking at an *unbroken continuum,* **one single continuous manifold of overlapping and compenetrated fields,** a kind of plasma, whose dynamic *intra*-relationships and valences account for every last feature of our Universe as we know it, from time and space to the diaphanous complexity of our human intelligence.[217]

The "nature" of *existence* — what it *is* — is to be seen in what it *does*. And what it does is to perdure as itself. *The "nature" of existence is to exist*. It evolves into myriads of forms and simultaneously "creates" time and space even as it remains itself.[218] *Matter's energy* remains itself through a process of sequential interior re-arrangement, an unfolding that has a communitarian character: the progressive elaboration of *integrated functions*.

Matter's energy is never found by itself in an unintegrated or uncombined state. It is intrinsically communitarian, creating bound relationships within itself, the better to survive. It is a dynamic self-possession built on and issuing in temporary stasis and endless change, as one tentative arrangement after another is used and transcended, searching intensely for a secure foothold in *existence*. All this change is simply the recombination of the selfsame substrate. It explains how *matter's energy* has developed into everything that *is-here* including the spacetime "envelope" in which "things" appear to exist. It displays itself as an endless dance of internal self-exploration — a self-unfolding that is at the same time a self-embrace. It is as if it were a single living organism.

---

[217] David Bohm, *Wholeness and the Implicate Order,* Routledge, London, 1989, p.220f. see appendix 5.

[218] Actually, it's more accurate to say it "*is* time and space," because spacetime does not exist apart from *matter's energy* under any conditions.

the human being — time and death

Existence is time.[219] It's not coincidental that *time* caused us to look at *being-here* separately and ask what it otherwise would not have occurred to us to ask, why do I die, or "*Why does being-here seem to end*?"

My life is both temporal *and* temporary. There's a connection between the two. It seems the very nature of the modulations of *existence* is to find *better* ways to *be-here,* to survive and extend survival. The vitality displayed by *matter's energy* is not a leisured aesthetic creativity, an unhurried pastime. There is an urgency here that derives from a **conatus**, a drive to survive, that is integral to a developing *universal entropy* that results from the energy expenditure of any "thing," whether it be the hydrogen fusing into helium in stars or the activity of the human brain. Entropy is the exhaust from combustion — the smoke that is the sign of fire — the tendency for all matter and energy in the universe to evolve toward a state of uniform inertia through the expenditure of energy for the performance of work. The aggregation and integration forged by *matter's energy* is part and parcel of the "downhill" flow of the existential cataract initiated at the big-bang that drives the Universe to produce its effects — like the eddies and vortices that spin off in a raging current. These pyramidal vortices (one vortex cumulatively building on another and another) are an anti-entropic phenomenon — they struggle against dissolution — even though they add to universal entropy as a result.

My life is the inner force of *existence* because it is *matter's energy.* It is driven in the direction of perdurance in an obsession to continue the dance of *presence.* Time is the effluence of my own *presence.* As my *existence* perdures from moment to moment it exudes time as the sweat of its creative labors; the vapor trail of its endless explorations. I embrace my *being-here,* and so I embrace time.

---

[219] The similarity of this proposition to Heidegger's thesis expounded in his *Being and Time* is only semantic. For H. time is the pulse and measure of Da-sein's anguish of being-toward-death, which alone brings Da-sein's *authentic care* to bear on the beings-in-the-world. In my conception, on the other hand, I make every effort to exclude the subjective factors. Time for me is foundationally a physical property exuded by the physical perdurance of a physical entity — *matter's energy.*

The transcendence over death, not only through evolutionary integration but also with other communitarian strategies like daily alimentation and organismic reproduction, harnesses even as it recapitulates the patterns and primordial energies let loose within the first second of the big bang. The energy that drives my hunger for existence, is the energy of matter itself.

We live in a banquet of *existence*. We are not only fed our own continued *presence* at this banquet, we also feed others.' In our lifetime, each human organism receives in sustenance probably 40 or 50 tons of the *matter's energy* of other living things who must die in order that we might live. And we prepare for our own opportunity to return our "stuff" to be used as food by others as part of an endless cycle of interchange within the one organism produced and energized by the cascade of *existence*.

At a certain magical moment, also, the very cells of my body, by utilizing another communitarian tactic, combine with another's to create a new identity — my daughter, my son — which is automatically granted a full allotment of time, slipping under the entropic radar of death. How was this miracle accomplished? The cells are mine, but their age and accumulated karma are erased. Death is cheated, fooled, outwitted. The new individual with my cells, my DNA, eludes the death they were otherwise destined to endure. Do we share this adventure in survival with love and gratitude?
Only if we understand!

But if we *mis*-understand — if we originally *mis*-interpreted that moment of crisis, the perception of death, as the cessation of what's really there, we are quite capable of turning this banquet of sharing into a selfish grab-bag where the desperate "eat drink and make merry" in a display of bitter disillusionment against a morrow of imagined nothingness. It is precisely the fact that "I" am metaphysically *in*significant except as an integrated function of matter's energy that opens me to a new dimension. I realize that what is really there and really important is the universal "stuff" of which I am made, the homogeneous substrate of which *all things* are made, the *single organism* of which we are all the leaves and branches, and which will go on in other forms endlessly. It was with those micro-threads of *existence* that I was woven. The primacy here, as always, belongs to the stuff of *existence*, the matter-energy of the universe. It is *material energy* "doing" me. And in short order, the same *existence* will use "me" to do something else in a constant search for "better" survival.

So time is the expression of process; it is the measure of groping and the tracks of creativity. It marks the work in progress of evolutionary development.

endless or "eternal"

The re-cycling is *endless*. Isn't that the same as "eternal," and doesn't it imply transcendent, necessary, absolute etc., all those characteristics derived from the "concept of being" that we rejected in chapter 1?

No. *Endless* is not "eternal" because endless is open and *empty*. "Eternal" is closed, fixed and finished, full and complete; "eternal" is the absence of time. *Endless,* on the other hand, *is* time ... time without end; it contemplates development without term. "Eternal," is synonymous with unchanging, impassible and immutable, Pure Act, pure *stasis*, without a shred of unfulfilled potential — perfect. It's a completely foreign concept to us, pure conceptual projection. We've never experienced anything the least bit like it. For us, *being-here* as we know it is an endless phenomenon that throbs always with unrealized potential, with an ever perceived emptiness seeking to be filled and asking for nothing *but more time*. We have never encountered *existence* in any other form. Its current modality is always in the process of becoming, apparently without limit, itself — *existence*.

Being-here in our world, is endless becoming. It's all we know. Where, then, do we get the notion of a fixed and finished "eternal"? I believe it's another of our fantasies based on the requirements of the imaginary ancient "concept of being." Existence, matter's energy, as found in the real world, however, is a function of *power* — as Spinoza discerned insightfully, *potentia* — *potential*; it is focused on *survival* and constantly ready to change tactics in order to achieve it. *Matter's creative power* is the drive *to exist* (*survive*) by extruding new forms out of itself in the context of time.

"Eternal" is unthinkable. *Endless* is not. We can understand *endless* perfectly because it's no different from time itself. To conceptualize "endless" requires no more insight than imagining present moments, "nows" in an open-ended flow into the future. In our very own awareness of ourselves-existing, which is the unfolding of our personal *presence* in time, we actually experience this phenomenon most intimately as *our own sentient selves*. We experience ourselves in a temporal flow into a potentially endless future. To experience temporal flow is to experience that part of "endless" which will always be here — the present moment, the only part of "endless" that ever … and always, exists. To experience one's own *presence in the here and now* is to experience, in a sense, **everything**, because it is to experience all that reality is, or ever was, or can ever be.

We are reminded that for the 14th century mystic Johannes Eckhart, "now" was the most sacred of all locations, the center of the universe. It was precisely where "God," he said, who exists in an Eternal "Now," was actively sharing "being" with creation in an effluence of love and self-donation. If you want to touch "him," he said, you can only do it "now." The fact that "now" is the only moment that *really exists* and that, at the same time, it goes almost universally unattended, may be a measure of exactly how alienated from *existence* we really are.

Can we say that our conception corresponds to the emphasis on living in the *present moment* promoted by the Buddhist, Thich Nat Hanh? The Buddhists insist their counsel is a discipline not a doctrine. They don't speak about metaphysics, "being" or *existence,* so we can't say for sure. But for the Buddhists, as for Meister Eckhart, *the present moment is all there is*. We *are-here* only in the present moment. To live in the present moment is to embrace the impermanence, the "emptiness" that drives reality always to the next moment.

being-here and emptiness (I)

The perdurance of *existence* in time is predicated on forging ever new relationships through combination, dissolution and re-combination — change and *movement* intended to satisfy what appears to be an inexplicable *need for existence*. What *presence* does is to tap its own potential for continued *presence*. Potential for *existence* can be said to be an *emptiness of presence* that seeks to be filled.[220] Hence its creativity. The thrust of its energies is always directed toward more secure ways of *being-here*. But, we have to ask, if *existence* is-here-now and *is-here* endlessly, how is it that the goal of its quest is still *existence*?

---

[220] **"emptiness"** — I use the word as a metaphor for the **conatus,** or drive to survive. I characterize it as a "hunger" or a "thirst" for *existence*. It's something we experience with varying degrees of intensity and "realization" throughout our lives.   The term is central to the vision of Nagárjuna, a 2nd century (ce) Mahayana Buddhist for whom "emptiness" refers to the fact that we do not *possess existence independently* but are rather "empty" of existence because we are dependent on other causes for our being here. For the Thomist tradition, with its emphasis on the *ontological dependency* of all things on **esse in se subsistens,** the same meaning is broadened and deepened.

It is the very restless instability of *being-here*, it's apparent radical insecurity, its resistance to the entropy that is its destiny, that appears to be *the source* of the endless energy of its explorations. *Existence* is not entirely reconciled to its fate. This characteristic of *existence* may have eluded identification when found in primitive, pre-life forms, but it reveals itself with indisputable clarity in living things. Life, as a manifestation of *matter's energy*, proves that *existence* is a mad desire, disruptive, violent, implacable.[221] The creativity of *being-here* is not a serene contemplative appreciation or a leisured aesthetic browsing. It is a passionate craving, an existential fury that seems to have no end.

Matter's energy is the *locus* of this insatiability. We say that because we see it functioning across the board. The frenzy of the oak tree to reach the sun, pathetic as it might appear, is not unfamiliar to us. We do the same in our own way as does every living thing that we know. The universality of the phenomenon of a generalized **existential hunger** that becomes growth, accumulation and self-aggrandizement, and I contend, evolutionary development, reveals to us the inherent qualities of *matter's energy* of which we are all made. Understanding that the qualities of life are due to its sub-atomic constituents, explains why insatiability, and from there, dissatisfaction, desire, anguish, obsession — suffering — is the lot of all organic life made from this universal, primordial clay. Humans are not exempt. Suffering, the sentient side of **emptiness,** cannot be ignored or assuaged. Any relief proves to be only temporary. It is endemic not only to *life,* but, we conclude, to *matter's energy itself.* To be is to live; to live is to suffer the throes of surviving. To survive — to stay the same — is to change, evolve, develop, complexify. It is to *create* out of emptiness *a world teeming with life.*

Life reveals reality. Once given the extended range of possibilities offered by that particular re-arrangement of matter's energy we call "life," it appears that *existence* flies its true colors. *Presence* is passionately and ruthlessly self-involved. Our praise for nature's exquisite balance cannot fail to recognize that this balance is achieved by an almost universal *violent predation*, as one species survives by heartlessly taking the life of another in order to incorporate its vital organic structures into its own. Predatory activity across the board is the basic tool of the natural system. In most cases it appears that evolutionary speciation — the very design of

---

[221] This approximates Schopenhauer's proposal that being is "will."

species — is a response to available prey, euphemistically called a "niche," or a "food source." Thus "nature" implants its blind lust for life, and seems impervious to the slaughter it engenders. On the one hand, this points up the unity and homogeneity of all material reality, for in fact one "entity" serves to support another. On the other, *hunger,* not only a metaphor in this case, appears to direct the process. Naturally the metaphors we use are themselves human as is the apparatus and the model, which is ourselves. *Emptiness, hunger,* are words that refer to human feelings that correspond to need. I don't apologize for this use of words. We can't escape from the fact the we, too, *are-here;* we survive by violence and we understand ourselves intimately by an *an understanding,* that recognizes that what constitutes us is our **implacable conatus**.

We saw in chapter 3 that certain activities, like self-replication and aggregation, once considered the exclusive domain of living things are also characteristic of non-living entities. We go even further and say that the very physical dynamisms operating in inanimate energy's relationships — gravitation, the strong and weak forces inside the atom, electromagnetism, chemical valences and molecular attraction — are actually constitutive elements of *matter's energy* as it aggregates, forming bound relationships, the better to survive. Words like "life" and "survive" are *metaphors* for pre-life integration taken from a resemblance to living things and human experience. But I claim they represent something *real* in the most fundamental forms of matter. We have identified that energy as *the **conatus***, the self-embrace of *existence,* a dynamism that uses similar strategies in response to an existential lack that characterizes all of *matter's energy.*

Lack? I believe we have touched a raw nerve in the organism of universal reality, an existential scar of such proportions that we are justified in calling matter's endless energy a function of **emptiness**. We understand the *conatus* as a wound of emptiness, because we understand ourselves.

east and west

While diverse cultures may agree on how to describe "emptiness," they have interpreted it variously and responded to it in different ways. In the West, following the belief in the transcendent importance of the individual person, *need* is identified as an obstacle to achievement, "self-transcendence." "Need" becomes a challenge — something to be overcome.

Emptiness, therefore, as an inherent and permanent defining factor integrates only as *antagonist* to "self-transcendence. "

In the East, on the other hand, Buddhists have a different take. *Emptiness,* they say, is *constitutive* of reality. Denying it is fatal and can be considered symptomatic of the human problem. Denial implies succumbing to the illusion of the possible permanence of the experiencing "self" and thus intensifies suffering. Buddhists believe that the false understanding of what the "self" really is (ultimately based on a mis-interpretation of what *existence* really is), encourages us to believe that we can somehow eliminate emptiness by engorging our "selves" with existence — meaning the accumulations that are thought to protect us against ultimate loss. That naturally includes wealth and power, and in our times, life-protecting technology. Religious practice as insurance for the after-life may be considered in this category. These accumulations promise to erase suffering, death, and ultimately permit us to live forever as our "*selves*" in another world.

The Buddhist view challenges these presuppositions. The hoarding, grasping selfishness created by the illusion that permanence can be achieved for the "self," only intensifies suffering for ourselves and everyone around us. What the *realization* called "enlightenment" does, they say, is to "awaken" us from the dream of permanence and to what is *really real.* From the point of view espoused in our reflections here, understanding reality to be *matter's energy* permits us to recognize that the *permanent self is an illusion*, that the craving and desire for this permanence is an unavoidable natural deception born of the internal dynamism of *matter's energy,* the emptiness which fuels the survival drive, that cannot be permanently satisfied. The implication is that we should understand emptiness as the ultimate definition of individuated reality. The appearance and increased complexification of the *integrated function* in the evolution of life is a direct product of the hungry emptiness that resides at the core of all reality, driving it to aggregate and integrate in order to avoid dissolution. Identity, then, which by reproduction creates species, is fundamentally an expression of existential need — *emptiness.*

The corollary to this Buddhist realization-awakening, one suspects, hovering in the background though officially unexpressed, is that what really exists and endures is the Whole of *being-here* taken as a Totality. It is the basis for the doctrine of *anatman*, the unreality of the "self." What

Buddhism claims to conquer is the aggravation of the cycle of suffering brought on by the mis-interpretation of what this "individual self" really is and therefore from the point of view of our reflections in this essay, what *being-here, existence,* with its endless *conatus* really is. We cannot escape suffering, they say, because we cannot escape from the *emptiness* and the consequent hunger for *existence* — the unreality — that resides at the core of things. Life ultimately cannot unseat death. Entropy wins.

Buddhism seems to suggest that to know reality is to understand the impermanence — the non-reality — of each and every feature and fact that emerges composed of *matter's energy* taken individually and apart from the Whole. Each individual manifestation of *presence* suffers from the same vulnerability because, at root, it cannot escape the primordial *emptiness* of its existential building blocks. The **conatus** characterizes all the strategies of survival and development as we saw. We are all made of the same "clay," and so, by ourselves, we all manifest the same characteristic impermanence that not only drives the communitarian strategy of *matter's energy* but also explains the clinging, grasping self-involved insecurity that causes so much of human suffering. The source of the energy at the base of the pyramid of reality is the *emptiness* inherent in any given separate manifestation of *being-here*. The *conatus* appears as if it were a reaction to an absence of *existence*.

But how can this be?

being-here and emptiness (II)

How can *existence* in any form, even partial, be *existentially empty*? If our analysis of *presence-as-process* is correct in saying that the fundamental dynamism of reality is *change and becoming*, and that change and becoming are in function of filling a need, then we find ourselves with an internal contradiction. Emptiness is *nothing*. As such it cannot be an explanation of the dynamism of *presence*.

If *existence* were simply static and at rest with itself, we would have no problem. But since *existence* displays itself as an endless becoming focused on being-here, "dragging" being-here into *existence* from moment to moment **as if it were not here at all,** we face a problem whose solution seems beyond the reach of our concepts. For as we perceive it, **existence acts as if it lacked the very thing that it is.** Lack of "being" can only mean non-being, "nothing." But, nothing, as we saw, is an absurd notion, because there is no such thing as "nothing." *Nothing* does not exist and therefore cannot be known. If it cannot be known, it cannot be conceptualized.

Existence, then, appears to be internally contradictory because by always moving to maintain itself it reveals an absence of self-possession. What is this absence? The circle of *presence* does not contain its explanation within itself. Where do we go from here? Beyond that circle, outside of *being-here*, human knowledge cannot function. For, outside of *existence,* there is nothing.

Haven't we gotten ourselves into this dead-end? After rejecting the validity of the traditional concept of "nothingness," haven't we simply resurrected it in another form, in a new guise, calling it *emptiness*? For what can *emptiness* "be" but another word for "nothingness?"

"emptiness" as metaphor

The impasse stated in this form is only apparent, and it arises from taking *emptiness* to be a "factual" or literal *concept* referring to "something" which can only mean "nothing." But emptiness is not nothingness because *emptiness* is not a concept, it is, as we've said all along, a *metaphor*. As metaphor, it does not answer, it rather preserves intact the significance of the *question.*

If we take *emptiness* as a literal concept and set *"presence" and "emptiness"* face-to-face, we discover that they cancel each other out; they cannot co-exist in the same mental construction. We cannot ask the question "how can *presence* be *empty*?" If "empty" is taken as a literal conventional concept, the question "how can presence be empty" is the same as asking "how can being be non-being." That contradiction means that we have no way of understanding reality. And I believe it's because we have confined our understanding of reality to what is mediated by conventional "literal" concepts and the so-called knowledge they produce. In the case we are considering that confinement is fatal. For "nothing" is a false concept, no matter what terms are used to describe it. It does not refer to anything at all.

Once we realize we are not using *emptiness* as a conventional concept, however, there is no inconsistency. *Emptiness* is a metaphor utilized to relate us to the dynamism of reality — reality's quest to remain itself. We have called it repeatedly, a self-embrace, and following Spinoza, **conatus**. Bergson called it the *vital impulse*,[222] Schopenhauer called it *will*.[223] In each case we are using an analogous human experience as *a metaphor* to describe this dynamism. We claimed we were justified in doing so because of the homogeneity of material reality. Everything is made of the same "stuff," *matter's energy*, including us. *Emptiness* does not refer to nothingness, but to a dynamism for self-possession, a self-embrace, which, when mediated exclusively by conceptual knowledge, is unintelligible. But, ironically, while we do not know what it is, when we approach it through our metaphors we realize that we do indeed *understand* it — intimately, thoroughly, profoundly, implicitly — because *we experience it as the inner dynamism of our very selves.* There is nothing in the world more familiar. It is our drive to survive. That is the basis for the validity of the metaphor.

It was otherwise with the traditional use of the abstract concept "nothingness," as we saw in chapter 1 and rejected. In that case there was an invalid attempt to generate a "proof" for the "necessity" of "being" based on the logical analysis of the opposition between the concepts of "being," taken literally, and "nothingness," also taken literally. Neither of those concepts was considered by the traditional metaphysicians to be anything but reliable representations of reality as it really is. It was precisely the impossible "reality" imputed to "nothingness," however, that gave us the first clue to the untenability of the entire procedure. The essentialists had reified the concept of "non-being" and then tried to make real inferences about the character of "being" from it.

Emptiness as we use it metaphorically, however, refers to an entirely different notion. Rationally speaking, the metaphor concretizes *the question* as a conceptual quest; it doesn't presume to provide a rational answer. We are proposing to *understand* the significance of an existential dynamic whose internal contradictions we cannot reconcile in conventional rationalist terms. The metaphor "emptiness," inspired by our bodily human experience and praeter-conceptual *understanding* of the phenomenon, describes in poetic terms what we do not conceptually comprehend but what we nevertheless experience and therefore *understand* intimately. This is a far cry from the claim to define the transcendent significance of "being" from a rational analysis of "non-being." Our use of the me-

---

[222] Cf *Creative Evolution*, 1907 passim
[223] Arthur Schopenhauer, *The World as Will and Idea*, Everyman London, 1995 tr Berman.

taphor "emptiness" immediately directs us to a recognition of the non-intelligibility of the concepts involved and from there to an acknowledged **conceptual ignorance**, even as it describes existence as we experience it with uncommon accuracy and undeniable certitude. Unlike the function of the concept "nothingness," which supposedly leads us to "know," *emptiness* (the metaphor) leads us to "not-know," or should we say to "*un-*know." *Emptiness* serves to put a human face on the baffling interior dynamism of all reality which we experience intimately as the very core of what we are. We understand it more clearly, more distinctly and more thoroughly than anything else in the world. And from there we *understand* all *existence.*

We *realize* that *existence* is *empty* for us because even though we *have* it, we still *thirst* for it — we know what that's like; we wake up with it everyday. But clearly it cannot be "known" in conventional conceptual terms, and therefore it cannot be controlled. We understand it, not because we conceptualize it or can identify its cause but because we experience it. We *realize* how accurately it defines us. It is a clear conscious embrace, a cognitively transparent experience but not a rational conceptual comprehension. *We understand it; but we do not know what it is.*

out of the impasse?

Rather than generate hypotheses to fill the conceptual gap, I am perfectly content that the final statement to be made on this question is that we can go no further — *conceptually.* We have encountered what Lonergan might have called a matter of sheer unintelligible *fact.*[224] The traditional "solutions" to the encounter with this philosophic dead-end, advanced in the West, in my opinion, have taken one of two paths. In the first, science-orientated reductionists ignore the problem by simply taking the existential dynamism for granted. They assume the unexplained existence of the embrace of *existence* and its manifestations in the survival drive and confine their analyses to what has subsequently evolved from it. They do not ask, as we do, what it is.

In the second, philosophers of the perennial essentialist tradition simply dismiss scientific questions as "not ultimate." They have no respect for *mere presence*, or "matter of fact."[225] They claim the real question exists

---

[224] For an extensive discussion of Lonergan's "unintelligible fact," see appendix 2.
[225] Cf. Rahner, *Spirit in the World.*, pp.162 and 175. And Lonergan, *Insight*, p.652.

only at the level of abstract "being" (and "non-being") and proceed to a "solution" by crediting our concepts and therefore the human mental apparatus with something they do not possess — a separate genus of being called "spirit." These "solutionists" (like Rahner and Lonergan) erect our very demands for *knowledge* into "proofs." Thus they continue the fundamental circularities that have characterized Western thought from the beginning. *I believe we have no justification for saying that the demand of our minds for an explanation is itself an explanation.* To my mind, this is to revisit the Platonic error and the Anselmian trap. We imagine reality based on the functions and products of our minds. To present human conceptual knowing (verbalized abstraction) in such a way that its description requires the *implied* existence of an unknown (and admittedly unknowable) object, is a huge projection.

Rahner says Thomas Aquinas agrees that human knowledge is locked into the limitations of sense experience. "Transcendence" by scholastic definition goes beyond those limits. So everyone agrees, including Thomas: *transcendence cannot be known directly.* Rahner's Thomas, however, is made to go further and say that the projections of human consciousness, (i.e., the ability to abstract), *imply* an absolute principle "pre-apprehended" by the mind, that never becomes itself the direct object of knowledge but opens us to another "realm" of knowledge. This is not a problem for Rahner because he believes "supernatural revelation" begins where direct knowing ends. The "absence of the implied object," in his system, plays a vital role in the transition to other "facts" in the form of revealed beliefs.[226]

At the end of my reflections the discovery of the *emptiness* at the heart of *being-here* puts me at a dead-end. I believe this is true of Spinoza, Schopenhauer and Bergson as well. I am aware that the apparent contradiction we encounter in the way matter's energy *is-here* leaves us at the edge of a void. We have reached the end of our earth-bound knowing. From a conceptual point of view, the rest is **darkness**. At that point Schopenhauer and Bergson each limit themselves to a description of that darkness — as "Will" or as "Vital impulse" — it's where the buck stops. Rahner, for his part, turns to revelation. What I claim, is that the only thing left ... if one has the temerity to go further ... is *relationship.*

---

[226] For a more complete treatment of this position see the appendix.

relationship to the darkness

In some way, then, that is not clear, we suspect that if there is an "explanation," it lies in that *darkness* into which we peer but cannot see — what we feel and touch as our very selves, what we *understand* so intimately and see so clearly and certainly but about which we can say nothing. We have little choice but to accept this situation because, however galling it might be, we ourselves awaken into a condition of absolute immersion in that darkness. We *understand* it with absolute clarity; we know of its creative power with absolute certainty; and we rely on it for our very *existence* itself. *Matter's energy*, the embrace of *existence*, is a matter of sheer unexplained empirical *fact*. It is as incomprehensible as it is absolutely familiar, undeniable and self-evident. It is the very fire and light of our lives, but utter darkness to our minds. It is us ... and yes indeed, we *understand it*.

What do I mean? If an *immersion-relationship* to *being-here* is the defining feature of our organisms, our *selves*, we fail to embrace the reliability of *existence* with its endemic thirst and *emptiness* at the risk of denying our very selves and the conditions under which we and our ancestors have *been here* and have evolved to become what we are. We cannot do that. We cannot sit in judgment on the circle of *existence, matter's energy, as if we stood outside of it;* for not only our faculty of analysis and judgment but our very *existence* itself is an evolved function of *matter's energy*. The internal incomprehensibility of *being-here* is now seen to have invaded our persons. The emptiness, the hunger to live, which we encountered in the dynamism of existence, material energy's self-embrace, we now see resides at the core of our very selves and lights the fire of our conscious presence; for we *are-here* without escape (not even death can annihilate the *material energy* that we are) and our very consciousness is a tool of our inherited determination to survive. We accept it. To fail to do so implies personal self-negation.

But notice: upon realizing that our analysis of *existence* could not explain itself, we did not physically annihilate nor disappear. Of course not. The contradictions we encountered in our rational ruminations had no impact whatsoever on *being-here. Existence* clearly is not dependent on our conceptualizations; the significance of *being-here* and the selectivity of rational consciousness do not move in the same plane. There is a reason why we cannot make deductions about reality from our ideas alone ... it's because our *understanding* of reality is not a function of ideas. Our consciousness is grounded in somatic experience, the organic immersion in *matter's energy*. It also supports our conclusion that the neo-Thomists' "transcendent thrust of consciousness" tells us nothing. Conceptualization with the logic of its required "explanations," in other words, does not correspond to the reality we have come to realize is *process* — energy, a dynamism we've described as a congenital self-embrace. And what we're interested in is *what reality is,* not how we conceptualize it.

The original organic function of abstractive intelligence was not "to know" but to *survive.* That we "do not know" is not a problem. It is the expression of the very nature of what we are. We were not meant to know; we were meant to survive. "Knowing" *what reality is,* is not an innate mission or mandate that comes from "God," as Rahner, Lonergan *et al.,* would have it. Knowing is a task we have set for ourselves. It's a valid project, but it's entirely ours; we cannot infer anything transcendent from our voluntary pursuit of it. Nor do we have a right to expect it will tell us what we demand: "knowledge" in terms of our warehoused ideas. Our inability to know is only a problem (or a solution, as for the Thomists) if we have assumed our conscious "selves" to be (as in fact we have in the West) like "gods," immortal spirits, striding above and beyond this world, forming divine immaterial ideas, the ultimate arbiters of all things material. We claim the right to sit in judgment on reality, submitting it to the bar of our dubiously reliable "ideas," as if our "raptor's claw" survival tool, abstractive conceptualization and its rationalist logic, were the very Mind of God.

In my opinion, this is the key. *We divinized human reasoning* — need I add, under the baneful influences of the Platonic-Cartesian illusions about the non-materiality of the human mind. From then on anything that does not yield to our concepts is judged irrational and impossible, all evidence to the contrary notwithstanding.

The evidence, however, does in fact withstand these presumptions. For, however absurd it may seem, we are-here ... and we understand it completely*!* Our being-here-now is something we cannot grasp with our rational intelligence, verbal-conceptual formulations and abstractive tools ... but that doesn't mean either that it is nothing or that we do not *understand* it. This reduces the range of possibilities offered by our conventional words even as it expands exponentially the potential for an accurate and intimate *understanding* of existence mediated by other cognitive mechanisms like metaphor, and the possibility of *relationship*. For our attempt to understand our conscious immersion in being-here translates to our attempt to understand the ineffable wordless darkness — that material energy with its existential self-embrace which we are.

"Darkness," of course, is another *metaphor* for this phenomenon, like *emptiness*. It is the living dynamism, the hunger of which we are constructed but unable to speak. It is what we are. In order to speak of this *immersion* we are forced to utilize our arsenal of non-conceptual apprehensions, our metaphorical allusions and poetic markers — myths, legends, parable-stories and witness personalities, rituals, symbols, interpretations and, most important of all, *contemplative silence*, to *evoke,* in a manner as close to *presence* itself as we can get, the embrace of *being-here* that we are. All we need do is experience ourselves *being-here* ... the rest follows.

Hence, at the end of the day, we realize we do not "know" ourselves, ... but we *understand* ourselves. We embrace ourselves in the transparent contemplation of a hungry and surviving energy that is "darkness" for our minds ... but only for our minds. It is an *understanding* of *existence* derived from the realizations and interpretations of what lies hidden in the crystalline clarity of *un*-knowing and the penetrating silence of interior experience. We *understand* this desire. It is who we are ... it is what everything is. It's why we understand one another ... and all things.

christian "revelation" and darkness

Christian "revelation," as traditionally understood and defended at least since the end of the middle ages, would turn this "darkness," this *un*-knowing, into "light," that is, into conventional *knowledge*. "Revelation," meaning beliefs, "factual truth" as we have inherited it, fundamentally claims to present clear ideas. It pretends to take the *emptiness and the darkness* out of *being-here* and to articulate it in the form of defined concepts provided by "divine authority" brokered exclusively by an infallible Church and/or the "Book." Catholic dogma is officially labeled *de fide definita* (a contradiction in terms, in my opinion). Dogma recapitulates the partializing distortions of abstraction that we have been trying to get in perspective throughout these reflections.

Conventional knowledge — concepts — is the unequivocal goal of Catholic dogmatic definitions. For, by claiming to "transcend" the dead-end of rational enquiry, "revelation" attempts to deny the ultimate significance of the unknowability, the *Mysterium Tremendum* that philosophy un-covered. The Void, the darkness, the emptiness, we must understand, is not a concept. It is the antithesis of all concepts. It is a *Mega-Metaphor*; the ultimate figure that describes our experience of being-here, our con-templative appreciation of the ineffable dynamism that drives becoming and gives meaning to our world and our very persons as part of that world. It is the force responsible for evolution. It is sacred for us for it is our very own lust for life. We experience it internally, we *understand* it intimately and with an incomparable certitude for it is ourselves, *but we do not know what it is.*

It's relevant to remember that before the Middle Ages, in the more an-cient Christian view, revelation was *not* considered defined dogma. Reve-lation for the ancients exclusively meant the Scriptures. John Scotus Eriúgena, for example, believed the result of rational enquiry, Philosophy, was *not transcended* by the Scriptures but rather was restated there in symbolic terms.[227] The Scriptures, he said, were allegories and symbols, "figures" (= metaphors) that represented the self-same truth discovered by Philosophy. We will recognize this as the view of all the Fathers from Ori-gen to Gregory of Nyssa in a living tradition that went back to Philo of Alexandria. In fact, for this tradition, as far as "knowledge of God" was concerned, Philosophy was the more direct and literal of the two. Scripture was believed to provide stories and symbols designed to make the ethe-real truths of Philosophy intelligible to the people who were not philoso-phers. The real "truth" contained in the symbols of scripture was Philo-sophical. Scripture did not trump Philosophy. The two were parallel modes of expression. There was only one "truth."

In this perspective, the bottomless Unknowable Ground into which the roots of reality sank and disappeared was a discovery of Philosophy that always remained insuperable. Ancient Christian mysticism as represen-ted by the *apophatic* tradition of Pseudo-Dionysius and Gregory of Nyssa, was constructed on exactly that foundation. Outside of the person and work of Jesus (who was quickly assimilated to Greek Philosophy's *Logos*), there was no "new" information about "God" to be found in the Scriptures. The Scriptures were symbols and stories which blended and flavored the "truth" of the Unfathomable Mystery — giving a "human" face to the Utter

---

[227] The end of the *Periphysion*

Darkness at the base of reality for the edification of the ordinary people. "God" was categorically unknowable and the role of revelation was only to provide metaphors for the darkness, not *knowledge.*

Since the days of the ascendancy of the claims of the infallibility of Catholic dogma, revelation has come to be presented not as figures and metaphors of the *unknowable,* but rather as "facts" that were allegedly *known* but just happened to be beyond unaided discovery and rational comprehension. This had a long historical development.[228] As the Church became associated with, and then progressively exercised in its own right the *imperial prerogatives* of the theocratic Roman State, its declarations about the "truth" became more arbitrary, authoritarian and "definitive." Beginning with Nicea (with the personal intervention of the Emperor Constantine himself), the Church acted as if it had inside information that defined "God," the Logos, the Trinity, Grace, the after-life, and was the only one that knew exactly how that information was to be used in practice. Fundamentally what it did was to reify *legitimate religious metaphors,* and turn them into gratuitously infallible dogmatic concepts, entities, qualities, reasons and explanations — *facts* taken literally. The upshot of this was to change the significance of mystery from "unknowable" to "unintelligible," and the method of expression from metaphor to defined dogmatic verbalized concept. As I grew up, every Catholic schoolchild was taught and believed that the "facts" of religion were fully known. The only "mystery" was what they meant*!*

But as far as "knowledge" was concerned, it meant that the Catholic Church "knew" everything that could possibly be known about "God." It solidified the Church's exclusive and universal role in "salvation." It was the basis for an ideological absolutism that dominated western culture for a thousand years and still has influence to this day.

preserve the question ... celebrate the darkness

The only way for religion to safeguard the integrity of the Unknown that our analysis of *presence-in-process* revealed to us, is to accept the "truths of revelation" not as conceptualized "facts" but as powerful evocative *metaphors,* creative instruments designed *to preserve the question,* not give an answer, ... to *celebrate un*-knowability, the "absent explanation," which is our life ... and to bundle the unknown remainder into *relationship* with

---

[228] This is similar to Adolph Harnack's assessment of the significance of Nicea as the first time that belief was accepted as irrational.

what, at root, is our very selves. For traditional Christianity this is not the 180° turn it appears to be. Our mystical traditions, going back past the Middle Ages, beyond the Cappadocian Fathers, beyond even Philo of Alexandria to the origins of Mosaic Yahwism, have always spoken of "God" as the Unknowable One. Moses' code demanded that graven images be forbidden lest we dared to imagine we "knew" the One-Who-Has-No-Name, *Yahweh*, which Philo tells us was a word that means "Nameless," "Imageless."[229] The surrender of the claim to possess conceptual "knowledge" of God means the end of "dogma." That will mean the *surrender of human control*, and an end to the arrogance of the sectarian religious enterprise. It accepts our ignorance. It confirms us in our utter humility, dethrones the overrated rational human "intellect" as the ultimate arbiter of reality, challenges the haughtiness spawned by our technological prowess and the false human superiority it implies, rejects the anti-material, anti-body, cerebral and gender-distorting assumptions of the Platonic-Cartesian Paradigm, and lays a solid foundation for ~~faith~~[230] not as arcane "knowledge," a canonical *gnosis,* but as unconditional trusting surrender to a darkness we embrace as the very core dynamism of our living selves.

I have intentionally used the same images and metaphors as the mystics, West and East, because I think we are talking about the same experience. Darkness, unknowing, emptiness, are traditional words that describe the fact that the only thing we will ever *know, conceptually,* is our universe of *matter's energy* — including us — endlessly driven to survive in the present moment.

To my mind, this is the basis for the ultimate reconciliation of philosophical enquiry and theological projection. It not only confirms the limited conclusions of rational observation and analysis at all levels, scientific and philosophical, but it also *guarantees* respect for the metaphors of all religious traditions which are attempting to celebrate and relate to the powerful creative darkness instead of denying it. It also finally includes in the circle of the *fully human* all those people branded "atheist," who choose to

---

[229] Philo of Alexandria, *On the Change of Names, II (7) to (14) passim,* tr.Yonge, Hendrickson Publishers, 1993, p.341-342.

[230] **faith:** I claim the word "faith" has been hijacked by its association with Christianity's projections about supernatural realities. Hence it is crossed out. That doesn't mean it's eliminated ... rather that it no longer has its traditional religious significance.

stand in utter silence before the mystery of it all, because they refuse to apply any metaphors whatsoever to the emptiness, the embrace of existence, that they, like the rest of us, encounter at the core of themselves. We are all made of the same thirsty clay, the same hungry quest for life. For those of us who know that the very heart of the matter is *that we do not know what that is*, "atheists" are our coreligionists.

But it should not make us disconsolate to say we do not *know*. We don't *need* to know; for we *understand* existence, and understanding opens to the possibility of *relationship*. Once we stop insisting that there must be an explanation that can be expressed in the conventional terms of our rational knowledge — concepts, explanations, reasons, words, logic, analyses, instruments of human control — the immense mystery of *being-here* discloses itself. For while we may not know what it is, we experience its dynamic power and understand it from within. We possess it completely in conscious form. For we are it. We can have no more intimate *understanding of* it than that. We can realize our identity with it; we can hold it and be-hold it in silent contemplation; and we can express, communicate and celebrate its groaning creative maternal benevolence which gave birth to this astonishing universe, with evocative metaphors, spellbinding myths and ecstatic rituals. And ultimately we love it as our very selves ...

But we do not know what it is.

Tony Equale
Willis, Virginia
2004-2009

appendices

appendix 1
aristotle's "metaphysics"

Metaphysics was once called a "science." Now we call it a *discipline*. It is a sophisticated intellectual exercise wedded to the perennial Western belief in the reality of "spirit." Metaphysics does not exist in a purely material world. The original Greek questioners 600 years before the Common Era were not metaphysicians, they were *cosmologists.* They ignored the "gods" and eschewed the existence of "spirit;" so, for them there was no metaphysics. Metaphysics did not exist until Aristotle catalogued the "sciences" three centuries later according to the dualist divisions he had internalized during his twenty years as Plato's associate.

For Aristotle the existence of *immaterial spirit* was a given; he neither discussed it nor felt he had to prove it. It was he who first separated out from physics the discipline later known as metaphysics which he called First Philosophy or Theology. He said that "theology" needed to be a separate science *not* because its subject was "God," the immutable source of all mutability, but because this "God" was an *immaterial being* and did not fall within the purview of physics whose proper object was material things.[231] First Philosophy was designed to deal with "being" in the sense of *everything,* the totality, and Aristotle's "being" included this incorporeal "God" and other "spiritual" realities.

Aristotle presents two separate analyses of "God's" creative relationship to the world as unmoved mover — in the "Physics" as efficient cause, and in the "Metaphysics" as final cause. Note that the notions of *immateriality* which implies mind (and therefore teleological purpose and "the Good") are all connected. First Philosophy had to be a separate science

---

[231] Richard McKeon, *Introduction to Aristotle*, Chicago U.Press, 1973, p.xxix.

because the operating principle — conscious *end or purpose* and the good intended ("final cause"), the exclusive products of "mind," therefore immaterial "spirit" — is entirely different from what's contemplated in the Physics. Metaphysics is the science of a "spirit," i.e., a "Mind" that acts with purpose.

First Philosophy came after "Physics" in Aristotle's collected works and so a posthumous editor called it exactly that: "what came after the physics" — *ta meta ta physica.* Hence, *metaphysics.* It stuck. In later times observers applied a figurative meaning to *meta,* "beyond," to announce that the subject of First Philosophy qualitatively transcended "Physics" whose subject matter was "only" the concrete reality of the physical world.

Aristotle's mediaeval commentators, Islamic, Jewish and Christian, identified "God" with "being" itself and so added the *abstract concept* of "being" onto the subject matter of metaphysics. It seems to have escaped their notice that by personifying "being" as "God" they made a substance of the term. This maneuver not only made "being" infinite and separated it from the only concrete existents in which it is ever found — the finite subject matter of Physics — it also meant the scholastics could only be talking about a mental construct, *an idea,* not a "real thing," thus ignoring the very basis of Aristotle's criticism of Plato.

Aristotle, remember, had denied there were independently existing ideas. The schoolmen may be forgiven for asserting *exactly that*, because they believed "being" was not just an idea, but a "real thing" called "God." And they could claim that it was Aristotle himself who led them there, for his theology was a mere rationalist revision of the imaginary *immaterial* "God" he had internalized as Plato's disciple. I see in this millennial myopia a confirmation of the cultural imperative requiring acceptance of the reality of "spirit" that dominated Western philosophical discourse from Plato onward. Aristotle, despite contrary assumptions, was its first obedient servant. That it was glaringly inconsistent with his own rejection of Plato's errors seems to have given him no pause.

But at first, before the Platonic era, there was no distinction between physics and metaphysics. The earliest philosophers ignored their puerile imaginary "gods," Apollo and Athena, Hera and Zeus. Reality, they said, was what you saw and touched in the real world. So to ask about reality was to ask about the cosmos: why things were *what* they were as well as *how and why* they were.   Thales, Anaximenes, Anaxagoras, Heraclitus, Parmenides, Democritus — they were all cosmo-ontologists.

Since I reject the *immaterial* assumptions that alone prompted Aristotle to certify *metaphysics* as a separate science, I also reject that certification. *Without a "spirit God" there is no metaphysics separate from cosmology.* The very distinction is a derivative of the Platonic Paradigm.   There is a separate metaphysics only in Plato's dual world of inert matter and substantial forms, living purposeful minds and dead random bodies, corrupt flesh and immortal spirit.

appendix 2
heidegger

heidegger, *das nichts* and being

In *An Introduction to Metaphysics* (lectures of 1935 published in 1953), Heidegger identifies the "being of beings" with the Greek concept of *phüsis* (φύσις), which he interprets to mean "power." "[*Phüsis*] is the power to emerge and endure."[232]   "The realm of being as such and as a whole is *phüsis*, i.e., its essence and character are defined as that which emerges and endures."[233]

The journey to that conclusion was not long or torturous.   Heidegger began with the same "metaphysical" question we examined in chapter 1 of the text, "why are there beings rather than nothing."[234]   Objections to using "nothing" as the constant in the equation were dismissed by Heidegger with short shrift, as one would expect, since *das Nichts* had become a central theme of his work, developed earlier in the short address entitled *What is Metaphysics* given in 1927.   There Heidegger says that *das*

---

[232] Heidegger, *Introduction...* Anchor 1962, p.12
[233] ibid., p.14
[234] ibid, p.1

Nichts, "nothingness," performs the horizon-function of disclosing the be-ing of beings.[235]

The definition of being as *phüsis* given in the *Introduction* flows inexor-ably from these premises. Heidegger says that the metaphysical ques-tion, *why is there something rather than nothing,* means,

> ... a ground is sought which will explain the emergence of the essent as an **overcoming of nothingness.** The ground that is now asked for is the ground of the **decision** for the present essent **over against nothingness**, ..."[236]

This "overcoming" and "decision" is the inevitable result of asking the "metaphysical question" in that particular way: *why is there something ra-ther than nothing* — as I have maintained, and criticized, in the text, chap-ter 1. For Heidegger, *das Nichts* — nothingness — is critical to the know-ledge of being. He says, "... [*das Nichts*] awakens for the first time the proper formulation of the metaphysical question concerning the being of beings. ... [it] reveals itself as belonging to the being of beings."[237] "Being and nothing belong together," Heidegger explains, "... because being is essentially finite and reveals itself only in the transcendence of Dasein which is held out into the [sic] nothing."[238] This critical connection between being and nothing is found also in Heidegger's study, *Kant and the Prob-lem of Metaphysics* (lectures of 1924; pub.1929). There, he starts from a statement of Kant's: "outside our knowledge we have nothing which we could set over against this knowledge as corresponding to it."[239] Heideg-ger then asks, How can we objectify the being of essents? How can a purely *receptive* faculty like consciousness generate the notion of "being" against an indistinguishable background of purely homogeneous "being?" What do we, on our own, use to objectify being? He answers,

> "It cannot be something essent [*because an essent is not distinguishable, as essent, from any other essent*]. If not an essent, then a Nothing [*Nichts*]. Only if the act of objectification is holding oneself into Nothing can an act of representation within this Nothing, let, in place of it, something not nothing, i.e., an essent, come forward to be met, ..."[240]

[235] *Basic Writings*, "What is Metaphysics," David Krell, ed., Harper Collins, NY, 1977, 1993, p.108.
[236] *Introduction* ...1956, p.23, emphases mine
[237] *op.cit*, Krell, p.108
[238] ibid.
[239] Immanuel Kant, *The Critique of Pure Reason*, A 104
[240] *Kant and the Problem* ..., 1929, tr Churchill, 1962, Indiana U. Press, p.76

So Heidegger says we generate the awareness of being against the background horizon of nothingness. Without such a background, we cannot apprehend "being" as a distinct object of knowledge.

But he goes further than Kant. He *reifies* the process. What for Kant perhaps, was an *a priori* mental operation, becomes for Heidegger a metaphysical dynamic: "being" emerges from and endures in the face of nothingness. Once being is seen as a conquest over nothingness, the notion of power follows inexorably. Being, then, is power, *phüsis*. Heidegger says of being-as-*phüsis:*

> [*phüsis*] denotes self-blossoming emergence (e.g., the blossoming of a rose), opening up, unfolding which manifests itself in such unfolding and preserves and endures in it; in short the realm of things that emerge and linger on. ... *phüsis* is being itself, by virtue of which essents [*beings*] become and remain observable. ... *Phüsis* means the power that emerges and the enduring realm under its sway. This power of emerging and enduring includes "becoming" as well as "being" in the restricted sense of inert duration. *Phüsis* is the process of a-rising, of emerging from the hidden whereby the hidden is first made to stand.[241]

By his insistence that being "emerges," Heidegger is clearly implying that it emerges from something *other than itself*. That can only mean from "nothing." This is confirmed when he contrasts this conception of *phüsis* with the use of the same word in a narrower "scientific" signification to refer to "nature," i.e., what is in motion of itself, *being emerging from being rather than from nothing*, including physical processes like the movements within the atom or the emergence of new forms of enduring being, like life. *Phüsis* in this more limited sense, according to Heidegger, is the proper subject of "physics" (we might add "biology"), not philosophy. Philosophical *phüsis* as originally conceived by the Greeks, he claims, was broader and deeper than "physics." He says it was beyond (*meta*) "physics" (*ta phüsika*) — hence, metaphysics. *Phüsis* included everything, he says; and it meant that everything, simply by existing, was as an expression of primordial power. Being was the power to emerge from nothing and the power to endure in the face of nothing.

Heidegger's disdain for the alleged degeneration of *phüsis* from its metaphysical meaning to its use in "physics" receives passionate expression. He declares that any claim that the Greeks actually began their musings with *phüsis* as "nature," would render them "primitives" ("Hoten-

---

[241] *Introduction* ... p.11-12

tots" was the word he used) and is an insult to their greatness and the greatness of Western thought.

Along with his contemptuous descriptions of this development as "decline" and "decay," he also makes derogatory reference to a very revealing corollary assumption, viz., if "*phüsis* is taken to be the fundamental manifestation of nature," he sneers, "then the first philosophy of the Greeks becomes a nature philosophy, *in which all things are held to be of a material nature.*"[242] He makes no further allusion to the issue, but what I hear in this snide aside are echoes of the Platonic Paradigm. There seems to be, for Heidegger, an unexpressed assumption that "being" *must* transcend the material form in which it is found (and what can that transcendence mean except the existence of *immaterial spirit* in some form or another.) Any contrary opinion merits neither argument nor rebuttal from Heidegger, only derision.

I believe this is the heart of the matter. The western assumption, accompanied by an invincible disdain for any other opinion, is that matter is dead, cold and inert; all signs of movement and vitality and the "power to emerge and endure" are due to "other" factors, which can only be immaterial "spirit" operating outside or alongside inert matter, as with Bergson's *vital impulse.* Otherwise, such vitality must be considered random, despite appearances. There is no way, according to this ancient prejudice, that it is even thinkable that *matter might possess within itself the dynamic source of its own presence* and the origins of everything that it subsequently shows itself capable of becoming. "Being" is the power to emerge and endure, but *matter.* as assumed in the West, is not.

matter is energy

Assumptions about the inertness of matter, however, have been demolished by the scientific discovery that *matter is energy.* That fact should have resulted in a complete overhaul in our thinking. And yet, it seems to have been no deterrence to the insistence that a metaphysical duality — *spirit and matter* — lies at the root of reality, thus maintaining the traditional "metaphysical problem." But if *being-here* is not split into spirit and matter the fruitless contortions of the metaphysicians to explain how the two can become one, are terminated and dismissed. The two are "one" to start with. Reality is, exactly as it appears: one homogeneous, inter-

---

[242] *Introduction*, op.cit., p.12 (emphasis mine).

connected and *intra*-related thing. Any search for what sustains *presence* must be sought *within the substrate itself,* by whose primordial dynamism (*phüsis*) all things "emerge and endure." The unimaginably slow and total- ly improbable emergence of an immense and *complex cosmos* out of the interactions within matter's energy are thus given an anticipatory focus: we *expect* to find therein the *meaning* of being — in terms that *we* can understand. For, as it stands now, if reduced to analyzing the random interactions of inert lifeless matter, or if confined to the endless autistic monolog of spirit-matter dualism, we *cannot understand cosmos* as we, its emergent children, experience it.

appendix 3
lonergan, rahner and "intelligible being"

Bernard Lonergan, like Karl Rahner, is an idealist in the Thomist tradi- tion. For both of them, their idealism stems from certain elements in the thought of Thomas Aquinas that he, in turn, inherited from the ancients. The first and foundational is the ancient Platonic belief in the independent super-reality of "ideas" responsible for a derived and shadowy reality in the "things" of the visible world. Scholasticism transposed Plato's *World of Ideas* into the creative "Mind of God." So, while Thomas' creation was no longer "shadows," as it was for Plato, "God" was still the Creator who thought finite "essences" reflective of the divine Essence and implanted them as "idea-forms" into matter, thus creating the universe of things that "participated" in the divinity of their Creator.

The second source of Thomas' idealism, defining both Lonergan and Rahner, is the doctrine of the "agent intellect." The agent intellect was an invention of Aristotle who had given it the more evocative name of *nous poiēticos* (from *poiein* "to make"). After rejecting Plato's theory of subsis- tent forms, Aristotle needed to account for the presence of universals in the human mind. He supposed an active function of the mind that is re- sponsible for abstraction. Following "the philosopher" then, Thomas taught that the intellect, as active, "extracts" the universal essence from the sense image by eliminating those (material) elements that belong to the particular concrete individual and identifying those which belong to the entire group. This produces an "intelligible species" which when "impressed" upon the "passive intellect" *(nous pathetikos)* results in the concept, i.e., understanding. Thomas calls the activity performed by the intellect "illumination." The active mind "illuminates" the essences embedded in the sense images. It frees things from their concrete particular material conditions revealing the "universal" essence.

Where does this light come from? According to Thomas the intellect shines *its own light* on the things it perceives and abstracts the universal essences from them. The intellect knows immaterial essences *in its own essence*. That implies that the human mind is in some kind of connatural relationship with the immateriality evident in all universals it can know — which is everything. *Aristotle claimed the active mind is eternal, pre-existed birth, and in some sense is a function of universal Nous which recalls Plato's "World Soul."* [243] Thus the active mind was connected to transcendence from its inception. Mediaeval philosophers like Roger Bacon believed the agent intellect was directly illuminated by God. [244]

Lonergan's version of the "agent intellect" is an "unlimited desire to know" which implies some kind of inchoate foreknowledge of transcendence. Rahner, for his part, calls it a "pre-apprehension" of infinity. Both men use it as a proof of the existence of transcendence, spirit ... and "God."

Thomist Idealism

In the view of Thomists, including Lonergan, real reality is only fully revealed insofar as it reflects and corroborates the products of the human mind. In other words, human mental phenomena — *ideas* — not only account for what's in our minds, ideas also *authentically reveal aspects of reality not available to perception.* Or, as the scholastics themselves might have put it, universal concepts touch on *the most real* aspect of the reality of things (i.e., the essence), an aspect that is not immediately available to sense perception. This is Thomistic idealism — what some call "modified realism." That's the basis of Lonergan's "intelligible being" which is not

---

[243] Aristotle, *De Anima*, Bk III, ch. 5, McKeon 2nd ed p.230

[244] "... he [Roger Bacon] comes to the conclusion in the *Opus Maius* that the greater philosophers in Greek, Islamic and Christian traditions maintained that the Agent Intellect *is* God, the source and agent of illumination." from the *Stanford Encyclopedia of Philosophy*, on-line. "Roger Bacon," 4.7

just a "matter of fact." "Intelligible being," for Lonergan, is a dense, loaded term pre-defined as "rational," and captured by the activity of the agent intellect in concepts that are "virtually unconditioned." Once his definition of "intelligible being" has been granted, the proof for the existence of God spins out of it with the speed of a bobsled.

Classic epistemological idealism turns on the assumptions regarding the transcendence of the universal concept over space and time. Matter was bound to space and time, so such transcendence could only be achieved by what was *immaterial*, i.e., spiritual. Classic (idealist) epistemology is intrinsically bound up with the belief in the existence of *spirit*. Indeed, Rahner's analysis of the "concept of being" in Thomas is pursued for the overt purpose of establishing human *spirituality*. It's not insignificant that his treatise is called *Spirit in the World.*

In contrast with all this, I define the concept as a simple "aggregation of past concrete experiences." (chapter 5). There is nothing "spiritual" in such a definition because we're only talking about an accumulation of concrete individualities, and a recognition of a simple plurality, not a universal created by an "insight" into an essence. Also, nothing is predicted about the future implying a grasp of what the thing is "in itself, always and everywhere." There is no "universal." The material explanation preserves the concept's provisional nature. And the traditional universal creates a problem in this regard. Let me explain.

The false universality of the traditional concept is an endemic problem that modern science must monitor constantly; it is a spontaneous error of common sense, *reinforced by the traditional epistemology*. To be more specific: science declares that every assertion of fact is tentative, dependent on continued verification in future experience. A definition is such an assertion, for science can and often does discover new characteristics of things that were not included in their original definitions. But in classic epistemology a definition (e.g., man is a rational animal) is the expression of a *concept that is universal* and infallibly true, therefore beyond any need for future verification. Lonergan himself deftly tries to dodge this clash by defining his "insight" in a way that avoids the infallible universalist claims traditionally made for *the concept*.[245] In the classic view the only

---

[245] Aristotle, *De Anima*, Bk III, Ch 6, McKeon, 2nd ed p.231-2 "In the thinking of the simple objects of thought, falsehood is impossible." and later, "the thinking of the definition, in the sense of the constitutive essence, is never in error."

source of possible error comes in the *judgment,* which may apply the infallible concept to the wrong entity. I may think a shadow at dusk is a man. That's an error; but my concept of "man" or humanity, is not. Lonergan, however, appears to want the "spirituality" implied in the activity of the agent intellect (with it's "unrestricted desire to know") without the "classic" consequences of asserting the (implied) universality of the resulting concept. He wants to make room for the valid empiricist insistence on the *provisional nature of* statements of fact. He limits all affirmations of universality to *judgments* (which *can* admit of error) that derive from a subsequent analysis performed upon the concept. But he can still claim the "universality" in the concept even though it is only *provisional.* He therefore calls the concept "*virtually* unconditioned."

This is important from his perspective. For if conceptualization does not transcend the material conditions of time and space, what claim does it have to "immateriality"? ... and consequently, what would that say about the alleged immateriality ("spirit") and transcendent reach of the agent intellect? The entire philosophical doctrine of human "spirit" is on the line in this highly finessed attempt to rescue a complicated explanation of how "matter" and "spirit" can possibly meet in knowledge.

the dilemma of dualism

This exemplifies the problem that plagues the classic world-view. When there is a dual world of spirit and matter, there is always the difficulty of explaining how two such diametrically opposed "principles of being" not only can co-exist, but interact, and constitutively, to form reality as we know it. Knowledge is a key point of contact between the two alleged worlds. But once conceded that there is no "spirit" as a separate genus of being, the need for these contorted epistemologies evaporates. If you took the "concept" as the word-labeled cumulative aggregation of similar concrete experiences (see chapter 5) ... then you no longer need to claim the concept is transcendent. Concepts are the tentative, linguistically labeled aggregations of like concrete experiences. They are not universal. They simply represent the recognition of similarity among many individual past experiences. There is nothing "transcendent" in that.

Animals also form concepts. Consider. The basis of the concept is the recognition of similarity. Animals that were incapable of recognizing similarity could never avoid predators because they would never *learn* that a second predator looked like, smelled like, sounded like, acted like an

earlier one. And if it is claimed that animals are not capable of this rudimentary conscious operation, this *learning,* then all such recognition
would have to be innately instinctual ... which is clearly contradicted by our
experience of the animals, especially of the higher domestic and farm animals that we are most familiar with, who clearly *learn.* The fundamental
recognition of the one and the many is the basis of all animal consciousness. It is not an exclusively human ability. It is not "spiritual."

I want to emphasize that my approach does not have to correct for
provisionality because it never claims universality to begin with. The concept is not a universal because it only refers to a composite of what the
past has offered, what has been concretely experienced; it doesn't pretend to reach an *essence* which necessarily predicts the future. The
epistemology I espouse agrees with science: it reduces all future predictions to probabilities.

from conceptual transcendence to metaphysical transcendence

In the next step, the Thomists go deeper. The more general basis of
the doctrine of the agent intellect is the Aristotelian belief that *the mind
must be of the same nature as the things it knows.* There must be a connaturality. Aristotle enunciates the principle in *De Anima*:

> "Interaction between two factors is held to require a precedent community of nature
> between the factors," Therefore, "... the mind will contain some element common to
> it with all other realities which makes them thinkable."[246]

Hence we have Aristotle's famous statement that "the mind is somehow all things," and from there the scholastic theorem that the "proper
object of the intellect is (infinite) being." The connaturality between the
spiritual mind and the material "thing," is had in the immaterial "essence,"
and the spirituality of the human mind bespeaks another connaturality between it and the Creating Spirit which is the Mind ("God") responsible for
the universe's immaterial essences. It's what makes "being" intelligible. It
is this corollary connection, made manifest in our "insight" into essences
as a function of our "unrestricted desire to know" that Lonergan uses to
ground our contact with "absolute transcendence." *"Intelligible being"* is a
phrase that sums up the entire vision.

Epistemological idealism begets ontological idealism. In Lonergan, the
opening to transcendence is displayed in a plethora of examples, de-

---

[246] Aristotle, *De Anima,* Bk III, Ch 4, #429: 24 ... in the 1973 revised McKeon edition, p. 229

signed to illuminate its deeper implications. And once again, the reasoning is thoroughly idealist: 1) I have "insight" driven by an "unrestricted desire to know *intelligible being*," 2) such a phenomenon necessarily implies an innate quest for infinite being, which means I somehow "pre-know" it. 3) the fore-knowledge of infinite being reveals finite reality as necessarily dependent on it, and that means that it must exist ... it goes from thought to reality. Anselm of Canterbury revisited. (See *Insight*, ch 19).

"intelligible" being

It all turns on the loaded notion of "*intelligible being*." Knowledge of intelligible being ("insight") is contrasted in Lonergan with mere "matter of fact." "Matter of fact" corresponds to what Rahner calls "mere presence." They are referring to what they believe is an empiricist over-simplification. The authors are proposing that there is a fatal lack of "reality" to things in themselves in the absence of the thinking subject actively having "insight" into their real realities (essences). The real realities are only revealed by *mind* ... which proves that those real realities are *intrinsically intelligible* ... *they are only real* because some creating Master Mind put them there with an (embedded) discernible purpose (essence). Without the *idea*-component, things are not "things," they are only "matter of fact." They are "mere presence." They are not "intelligible." They are not *real*. They are "nothing." Hear them speak: First Lonergan:

> ... being is intelligible ... it is what is to be known by intelligent grasp and reasonable affirmation. ... what is apart from being is nothing, so what is apart from intelligibility is nothing. It follows that to talk about mere matters of fact that admit no explanation is to talk about nothing. If existence is mere matter of fact, it is nothing. If occurrence is mere matter of fact, it is nothing. If it is mere matter of fact that we know and that there are to be known classical and statistical laws, generic operators and their dialectical perturbations, explanatory genera and species, emergent probability and upward finalistic dynamism, then both the knowing and the known are nothing.[247]

He's saying that knowledge is insight *as he's defined it,* and to claim to "know" things without it (which he calls "matter of fact") is a contradiction in terms. Such a claim, furthermore, can itself only be made on the basis of having "insight." So from his perspective, every attempt to refute his assertion involves the proof of his assertion.

[247] *Insight*, p.652

Now here's Rahner on the same issue — what he calls "mere presence," a phrase that corresponds to Lonergan's "matter of fact": (emphasis mine):

> But that [if *esse* is not to be nothing] requires a radical revision of the common concept of *esse*. *Esse* is no longer *mere presence*, the indifferent ground, as it were, upon which identical and undifferentiated ground the different essences must stand, if in addition to their real ideal being they also wish to be reality. ... *esse* does not mean *the empty indifference of a mere existence* which prescinds completely from what exists by it. [248]

"*What* exists by it," of course are "things," in Lonergan's terms, and things (erstwhile classical "substances") are *what* they are by their "*whatness*," their "*quiddity*," their essence. Lonergan sums it up unequivocally:

> "Hence being apart from essence is being apart from the possibility of existence; it is being that cannot exist; but what cannot exist is nothing, and so the notion of being apart from essence is the notion of nothing." [249]

So "being" is pre-loaded as "intelligible," and "rational." Hence every instance of "proportionate being" (finite, contingent being) always implies a ground, an explanation. But that's because Lonergan has pre-defined it that way. And he defined it that way because he examined a notion of "being" that was pre-formed in his mind by a world-view that presumed that universals are "essences" and constitute understanding. He started from a prejudiced "insight." The circularity here is systemic.

a groundless ground

But notice. In Lonergan's system, the *ultimate ground* that all contingent being implies *itself has no ground*. It is self-grounded. As such, by the definitions adduced to confound the empiricists, it's supposed to be "unintelligible" because it has no explanation. So, to say that as the anchor of the chain of causes *it needs no explanation* because it is its own explanation is to tacitly admit that we "connaturally" understand that which needs no explanation. But that which has no explanation, by Lonergan's definition, is *only a matter of fact*. It is nothing ... and supposedly unthinkable, unintelligible. So at the end of the chain that spins out from insight, is an insight into *sheer matter of fact* ... which was earlier defined as impossible.

---

[248] *Spirit in the World*, p,162, 175-6
[249] *Insight*, pp. 371-2.

But from my point of view, if you examined *existence* from what actually *exists,* without any essentialist presumptions, you discover empirically that *what* things are, in fact, is an *ad hoc* product, an adventitious modality of a homogeneous substrate (*material energy*) arrived at not rationally (purposefully) but opportunistically — by chance — through the hysterical, mindless, irrational groping driven by the "senseless" insistence on surviving, i.e., *existing, the* **conatus sese conservandi.** Here's the question: What explains why things ferociously insist on *existing ... on being here*? ... and therefore why they originate species? Answer: Nothing whatsoever! It has no explanation beyond itself. It has *no purpose* beyond itself. **The purpose of existence is to exist.** There is nothing besides *existence* to explain why things are driven to *continue to exist. Existence is a congenital self-embrace. It's an inexplicable matter of fact.* And it is this "matter of fact," as a matter of fact, that makes things *what* they are. For the drive to survive is precisely what *originates species.* It is creative. It constitutes all things.

So then, what do we say ... that because it has no "rational" explanation beyond itself that *existence is unintelligible* ... that we don't "understand" *existence*? Not in the least. We understand *existence* perfectly, intimately, connaturally, internally, implicitly *for we ourselves exist and we understand the "drive to survive," from within. We understand the "self embrace" like nothing else in the world. It constitutes us. It's who and what we are. We know exactly what it means to exist and to want desperately to continue to exist and that we will virtually stop at nothing to keep on existing ... and that every blessed thing out there is doing exactly the same thing as we are.* If there's one thing in this world that we *understand* thoroughly, "as a matter of fact," it's *existence.* We *understand* it directly in itself, not "as explained" in and through something else.

It's the initial idealist definitions given to the "concept of being" that cast the traditional argument in steel. "Being" for Lonergan is only what the rational mind can make it — intelligible, explained, the product of "insight."

But without "essences" there is no "intelligible being." "Intelligible being" is necessarily the product of the confluence of spirit and matter, a metaphysical dualism. In the radical empiricist position, however, there are no dualisms to be reconciled and hence no "intelligible being" is required. There is only one thing out there: *existence, matter's energy.* That's what exists. And if that's what exists, then it's "being," and therefore it's the benevolent source of our *existence* — "God," if you will. *Material energy* generates a sense of the Sacred in us, because it's what we are made of, it gave us ourselves. We love being us and being here. There is no alternative.

Unthinkable? ... counter-intuitive for sure, given our tradition. But please note: this substrate is *not* the inert "matter" our ancient dualist tradition thought it saw and trained us to see. We have taken another look and we can see for ourselves that *matter's energy* is *not* dead passivity, and it is *not* reducible to mere billiard-ball interactions. Those were prejudicial Cartesian mis-characterizations generated by a world view dominated by a belief in "spirit." *Matter's energy* in the real world is alive, *a la* Bergson and Schopenhauer, and developing, *a la* De Chardin and Whitehead. And the story of its anguished creativity is engraved in the keening synclines of our planet's tortured crust, and in the desperate detours recorded in the convoluted helical codes of life's DNA. Despite all its tormented groping and *endemic suffering* and perennial failures, it is ongoing. It will not stop anywhere, and it stops at nothing. And we *understand* it well ... all too well ... for we are it.

appendix 4:
conatus

From Wikipedia, the free encyclopedia

Conatus (Latin for *effort; endeavor; impulse, inclination, tendency; undertaking; striving*) is a term used in early philosophies of psychology and metaphysics to refer to an innate inclination of a thing to continue to exist and enhance itself. This "thing" may be mind, matter or a combination of both. Over the millennia, many different definitions and treatments have been formulated by philosophers. Seventeenth-century philosophers René Descartes, Baruch Spinoza, and Gottfried Leibniz, and their Empiricist contemporary Thomas Hobbes made important contributions. The *conatus* may refer to the instinctive "will to live" of animals or to various metaphysical theories of motion and inertia. Often the concept is associated with God's will in a pantheist view of Nature. The concept may be broken up into separate definitions for the mind and body and split when discussing centrifugal force and inertia.

The history of the term *conatus* is that of a series of subtle tweaks in meaning and clarifications of scope developed over the course of two and a half millennia. In adopting the term successive philosophers put their own personal twist on the concept, each developing the term differently such that it now has no concrete and universally accepted definition. The earliest authors to discuss *conatus* wrote primarily in Latin, basing their usage on ancient Greek concepts. The Latin *conatus* comes from the verb *conor*, which is usually translated into English as, "to endeavor"; but the concept of the *conatus* was first developed by the Stoics (333–264 BCE) and Peripatetics (c. 335 BCE) before the Common Era. These groups used the word ὁρμή (hormê, translated in Latin by *impetus*) to describe the movement of the soul towards an object, and from which a physical act results. These thinkers therefore used "*conatus*" not only as a technical term but as a common word and in a general sense. In archaic texts, the more technical usage is difficult to discern from the more common one, and they are also hard to differentiate in translation. In English translations, the term is italicized when used in the technical sense or translated and followed by *conatus* in brackets. Today, *conatus* is rarely used in the technical sense, since modern physics uses concepts such as inertia and conservation of momentum that have superseded it. It has, however, been a notable influence on nineteenth- and twentieth-century thinkers such as Arthur Schopenhauer, Friedrich Nietzsche, and Louis Dumont.

appendix 5:
the universe is an undivided whole

(a) from physicist-philosopher David Bohm:

Let us first consider the mechanistic order. The principal feature of this order is that the world is regarded as constituted of entities which are outside of each other, in the sense that they exist independently in different regions of space and time and interact through forces that do not bring about any changes in their essential natures. The machine gives a typical illustration of such a system of order. Each part is formed independently of the others and interacts with the other parts only through some kind of external contact. . . .

... physics has become almost totally committed to the notion that the order of the universe is basically mechanistic. The most common form of this notion is that the world is assumed to be constituted of a set of separately constituted, indivisible and unchangeable "elementary particles," which are the fundamental "building blocks" of the entire universe.

[But the theory of relativity] implied that no coherent concept of an independently existing particle is possible ... Einstein proposed that the particle concept no longer be taken as primary, and that instead reality be regarded from the very beginning as constituted of fields, obeying laws that are consistent with the requirements of the theory of relativity. A key new idea of this "unified field theory" of Einstein is that the field equations ... could have solutions in the form of localized pulses, consisting of a region of intense field that could move through space stably as a whole, and that could thus provide a model of the "particle." Such pulses do not end abruptly but spread out to arbitrarily large distances with decreasing intensity. Thus the field structures associated with two pulses [*erstwhile 'particles' ed.*] will merge and flow together in one unbroken whole. Moreover when two pulses come close together, the original particle-like forms will be so radically altered that there is no longer even a resemblance to a structure consisting of two particles. So, in terms of this notion, the idea of a separately and independently existent particle is seen to be, at best, an abstraction furnishing a valid approximation only in a certain limited domain. Ultimately, the entire universe (with all its "particles," including those constituting human beings, their laboratories, observing instruments, etc.) has to be understood as a single undivided whole, in which analysis into separately and independently existent parts has no fundamental status.

Wholeness and the Implicate Order, London, Routledge, 1980, p. 220f

(b) from physicist Brian Greene

The need to abandon locality is the most astonishing lesson arising from the work of Einstein, Podolsky, Rosen, Bohm, Bell and Aspect and many others ... . By virtue of their past, objects that at present are in vastly different regions of the universe can be part of a quantum mechanically entangled whole. Even though widely separated, such objects are committed to behaving in a random but coordinated manner.

   We used to think that a basic property of space is that it separates and distinguishes one object from another. But we now see that quantum mechanics radically challenges this view. Two things can be separated by an enormous amount of space and yet not have a fully independent existence. A quantum connection can unite them, making the properties of each contingent on the properties of the other. Space does not distinguish such entangled objects. Space cannot overcome their interconnection. Space, even a huge amount of space, does not weaken their quantum mechanical interdependence.

The Fabric of the Cosmos, NY Vintage, 2004, p.122

appendix 6:
the materiality of "God"

(a) the anomaly of an immaterial creator of a material universe

From Gregory of Nyssa (Fourth Century),
On the Making of Man, XXIII, 3 and 4.,

"3. ... they [who challenge the possibility of **matter** being created by an **immaterial God**] employ in support of their own doctrine some such arguments as these: If God is in His nature simple and immaterial, without quantity, or size, or combination, and removed from the idea of circumscription by way of figure, while all matter is apprehended in extension measured by intervals, and does not escape the apprehension of our senses, but becomes known to us in color, and figure, and bulk, and size, and resistance, and the other attributes belonging to it, none of which it is possible to conceive in the Divine nature, — what method is there for the production of matter from the immaterial, or of the nature that has dimensions from that which is unextended? For if these things are believed to have their existence from that source, they clearly come into existence after being in Him in some mysterious way; but if material existence was in Him, how can He be immaterial while including matter in Himself? And similarly with all the other marks by which the material nature is differentiated; if quantity exists in God, how is God without quantity? If the compound nature exists in Him, how is He simple, without parts and without combination? so that the argument forces us to think *either that He is material, because matter has its existence from Him as a source*; or, if one avoids this, it is necessary to suppose that matter was imported by Him *ab extra* for the making of the universe.

4. If, then, it [matter] was external to God, something else surely existed besides God, conceived, in respect of eternity, together with Him Who exists ungenerately; so that the argument supposes two eternal and unbegotten existences, having their being concurrently with each other — that of Him Who operates as an artificer, and that of the thing which admits this skilled operation; ... Yet we do believe that all things are of God, as **we hear the Scripture say so; and as to the question how they were in God, a question beyond our reason**, **we do not seek to pry into it**, believing that all things are within the capacity of God's power — both to give existence to what is not, and to implant qualities at His pleasure in what is.

(b) extension is an attribute of God

Baruch Spinoza
from the *Ethics*, 1677

Part II, prop 2. *Extension is an attribute of God; i.e., God is an extended thing.*

Part I, prop 15, Scholium: Some assert that God, like a man, consists of body and mind, and is susceptible of passions. How far such persons have strayed from the truth is sufficiently evident from what has been said. But these I pass over. For all who have in anywise reflected on the divine nature deny that God has a body. Of this they find excellent proof in the fact that we understand by body a definite quantity, so long, so broad, so deep, bounded by a certain shape, and it is the height of absurdity to predicate such a thing of God, a being absolutely infinite. But meanwhile by the other reasons with which they try to prove their point, they show that they think corporeal or extended substance wholly apart from the divine nature, and say it was created by God. Wherefrom the divine nature [as extension] can have been created, they are wholly ignorant; thus they clearly show, that they do not know the meaning of their own words. I myself have proved sufficiently clearly, at any rate in my own judgment, that *no substance can be produced or created by anything other than itself.* Further, I showed (in Prop. xiv.), that *besides God no substance can be granted or conceived.* Hence we drew the conclusion that *extended substance is one of the infinite attributes of God.* However, in order to explain more fully, I will refute the arguments of my adversaries, which all start from the following points:--

Extended substance, in so far as it is substance, consists, as they think, in parts, wherefore they deny that it can be infinite, or, consequently, that it can appertain to God. This they illustrate with many examples ... [and] try to prove that extended substance is unworthy of the divine nature, and cannot possibly appertain thereto. . . . Moreover, any one who reflects will see that all these absurdities, from which it is sought to extract the conclusion that extended substance is finite, do not at all follow from the notion of an infinite quantity, but merely from the notion that an infinite quantity is measurable, and composed of finite parts; therefore, the only fair conclusion to be drawn is that infinite quantity is not measurable, and cannot be composed of finite parts. This is exactly what we have already proved (in Prop. xii.). ... For, taking *extended substance, which can only be conceived as infinite, one, and indivisible* (Props. viii., v., xii.) they assert, in order to prove that it is finite, that it is composed of finite parts, and that it can be multiplied and divided.

So, also, others, after asserting that a line is composed of points, can produce many arguments to prove that a line cannot be infinitely divided. Assuredly it is not less absurd to assert that extended substance is made up of bodies or parts, than it would be to assert that a solid is made up of surfaces, a surface of lines, and a line of points. ... For if extended substance could be so divided that its parts were really separate, why should not one part admit of being destroyed, the others remaining joined together as before? And why should all be so fitted into one another as to leave no vacuum? Surely in the case of things, which are really distinct one from the other, one can exist without the other, and can remain in its original condition. As then, there does not exist a vacuum in nature, but all parts are bound to come together to prevent it, it follows from this also that *the parts cannot be really distinguished, and that extended substance in so far as it is substance cannot be divided.*

. . . If, then, we regard quantity as it is represented *in our imagination,* which we often and more easily do, we shall find that it is finite, divisible, and compounded of parts; but if we regard it as it is represented *in our intellect,* and conceive it as substance, which it is very difficult to do, we shall then, as I have sufficiently proved, find that it is infinite, one, and in-divisible. This will be plain enough to all, who make a distinction between the intellect and the imagination, especially if it be remembered, that matter is everywhere the same, that its parts are not distinguishable, except in so far as we conceive matter as diversely modified, whence its parts are distinguished, not really, but modally. . . .

I think I have now answered the second argument; it is, in fact, founded on the same assumption as the first — namely, that matter, in so far as it is substance, is divisible, and composed of parts. Even if it were so, I do not know why it should be considered unworthy of the divine na-ture, inasmuch as besides God (by Prop. xiv.) no substance can be gran-ted, wherefrom it could receive its modifications. *All things, I repeat, are in God,* and all things which come to pass, come to pass solely through the laws of the infinite nature of God, and follow (as I will shortly show) from the necessity of his essence. Wherefore it can in nowise be said, that God is passive in respect to anything other than himself, or that extended sub-stance is unworthy of the Divine nature, even if it be supposed divisible, so long as it is granted to be infinite and eternal.

appendix 7
emergence

(from **"emergence."** *Encyclopædia Britannica* 2006 Ultimate Reference Suite DVD)

Emergence, in evolutionary theory, means the rise of a system that cannot be predicted or explained from antecedent conditions. George Henry Lewes, the 19th-century English philosopher of science, distin-guished between resultants and emergents — phenomena that are pre-dictable from their constituent parts and those that are not (*e.g.,* a physical mixture of sand and talcum powder as contrasted with a chemical compound such as salt, which looks nothing like sodium or chlorine). The evolutionary account of life is a continuous history marked by stages at which fundamentally new forms have appeared: (1) the origin of life; (2) the origin of nucleus-bearing protozoa [eukaryotes]; (3) the origin of sexually reproducing forms, with an individual destiny lacking in cells that reproduce by fission; (4) the rise of sentient animals, with nervous systems and protobrains; and (5) the appearance of cogitative animals, namely humans. Each of these new modes of life, though grounded in the physicochemical and biochemical conditions of the previous and simpler stage, is intelligible only in terms of its own ordering principle. These are thus cases of **emergence**.

Early in the 20th century, the British zoologist C. Lloyd Morgan, one of the founders of animal psychology, emphasized the antipode of the principle: nothing should be called an emergent unless it can be shown not to be a resultant. Like Lewes, he treated the distinction as inductive and empirical, not as metempirical or metaphysical — *i.e.,* not beyond the observable realm. Morgan condemned the 20th-century French intuitionist Henri Bergson's creative evolution as speculative, while proclaiming emergent evolution as a scientific theory. Even so, the theory has not been accepted universally by biologists. With genetics illuminating the mechanism of heredity (and hence the very conditions of evolution) and biochemistry elucidating the workings of the cell nucleus, some biologists are confirmed in their belief that *scientific treatment admits only of analysis into parts and not into new kinds of wholes.* [emphasis mine]. Thus, they tend to concentrate on the mechanisms of mutation and of natural selection, effective in microevolution — the change from variety to variety and species to species — and to extrapolate these findings to macroevolution, to the origin of the great groups of living things.

Nevertheless, the concept of **emergence** still figures in some evolutionary thinking. In the 1920s and '30s, Samuel Alexander, a British realist metaphysician, and Jan Smuts, the South African statesman, espoused **emergence** theories; and later, others, such as the Jesuit paleontologist Pierre Teilhard de Chardin and the French zoologist Albert Vandel, emphasized the series of levels of organization, moving toward higher forms of consciousness. The philosophy of organism of Alfred North Whitehead, the leading process metaphysician, with its doctrine of creative advance, is a philosophy of **emergence**; so also is the theory of personal knowledge of Michael Polanyi, a Hungarian scientist and philosopher, with its levels of being and of knowing, none of which are wholly intelligible to those they describe.

Emergence

(from **"complexity."** *Encyclopaedia Britanica* 2006 Ultimate Reference Suite DVD)

A surprise-generating mechanism dependent on connectivity for its very existence is the phenomenon known as **emergence**. **Emergence** refers to unexpected global system properties, not present in any of the individual subsystems, which emerge from component interactions. A good example is water, whose distinguishing characteristics are its natural form as a liquid and its nonflammability — both of which are totally different from the properties of its component gases, hydrogen and oxygen.

The difference between complexity arising from **emergence** and that coming only from connection patterns lies in the nature of the interactions between the various components of the system. For **emergence**, attention is not placed simply on whether there is some kind of interaction between the components but also on the specific nature of those interactions. For instance, connectivity alone would not enable one to distinguish between ordinary tap water, which involves an interaction between hydrogen and oxygen molecules, and heavy water (deuterium), which involves an interaction between the same components but with an extra neutron thrown into the mix. **Emergence** would make this distinction. In practice it is often difficult (and unnecessary) to differentiate between connectivity and **emergence**, and they are frequently treated as synonymous surprise-generating mechanisms.

Emergence in an ant colony

Complex systems produce surprising behaviour; in fact, they produce behavioral patterns and properties that just cannot be predicted from knowledge of their parts taken in isolation. The appearance of emergent properties is probably the single most distinguishing feature of complex systems. … An example of **emergence** occurs in the global behaviour of an ant colony.

Like human societies, ant colonies achieve things that no individual member can accomplish. Nests are erected and maintained; chambers and tunnels are excavated; and territories are defended. Individual ants acting in accord with simple, local information carry on all of these activities; *there is no master ant overseeing the entire colony and broadcasting instructions to the individual workers.* Each individual ant processes the partial information available to it in order to decide which of the many possible functional roles it should play in the colony.

Recent work on harvester ants[250] has shed considerable light on the processes by which members of an ant colony assume various roles. These studies identify four distinct tasks that an adult harvester-ant worker can perform outside the nest: foraging, patrolling, nest maintenance, and midden work (building and sorting the colony's refuse pile). It is primarily the interactions between ants performing these tasks that give rise to emergent phenomena in the ant colony.

When debris is piled near their nest opening, nest-maintenance workers abound. Apparently, the ants engage in task switching, by which the local decision of each individual ant determines much of the coordinated behaviour of the entire colony. Task allocation depends on two kinds of decisions made by individual ants. First, there is the decision about which task to perform, followed by the decision of whether to be active in this task. As already noted, these decisions are based solely on local information; there is no centralized control keeping track of the big picture.

Once an ant becomes a forager it never switches to other tasks outside the nest. When a large cleaning chore arises on the surface of the nest, new nest-maintenance workers are recruited from ants working inside the nest, not from workers performing tasks on the outside. When there is a disturbance, such as an intrusion by foreign ants, nest-maintenance workers switch tasks to become patrollers. Finally, once an ant is allocated a task outside the nest, it never returns to chores on the inside.

The foregoing ant colony example shows how interactions between various types of ants can give rise to patterns of global work allocation in the colony, emergent patterns that cannot be predicted or that cannot even arise for isolated ants.

---

[250] Deborah Gordon, *Ants at Work*, 1999, WW Norton, NY

appendix 8
spinoza

It hardly needs to be said that there is a great deal of similarity between Spinoza's vision and that of the *Mystery of Matter.*

Both philosophies are ***monist***; both propose that the **materiality** of the universe is not derivative and conditioned but primordial, causative and originating; both affirm that the universe of things is the result of a **self-unfolding in time** that correlates with a ***conatus*** or **self-embrace.**

But at the same time the differences are not insignificant.

* Spinoza's vision is **idealist**. It ascribes reality and ontological power to *ideas* whose interrelationships it then tries to schematize. Spinoza's geometric method is essential to the structure of that schema as well as for didactic purposes and "scientific" proof.

   ... *MM, on the other hand,* is an inductive **empiricism** that corresponds to the quantum and evolutionary science of the 21st century; it identifies *material energy in process* as the single factor that both creates and explains existence in all its forms and functions.

* Spinoza's single substance is **God**, known by an allegedly self-evident proposition. This "God" is the immanent cause and explanation of all things in *what* as well as *that* they are. *What* things are (essences) spring from God as a thinking thing, and take on bodies that spring from God as an extended thing.

   ... *MM's* single substance, on the other hand, is a substrate called **material energy** with a connatural self-embrace (***conatus***) that unfolds to become all things. This material energy with its *conatus* **may** be infinite, necessary and necessarily creative (and therefore "God,") but there is no way any of that can be declared certain.

* Spinoza distinguishes between "**extension**" and "**thought**" as two equal, separate and distinct principles of being ... thus introducing a potential[251] **dualism** at a second level (below "God"). Despite their ir-

---

[251] I say "potential" here because as applied by Spinoza to the nature of mind in Part II of the *Ethics*, there is a clear attempt to subordinate mind to body.

reducibility to one another; however, he claims they are simply aspects of one and the same thing — they are "attributes" of God.

... *MM* states simply that consciousness is a property of material energy whose potential for thought evolves through the *conatus*.

*   Spinoza uses a **deductive methodolog**y insisting it has a corroborative value that supports the accuracy and coherence of his system;

... *MM* works exclusively through **induction and interpretation** to arrive at hypotheses that need verification.

*   Spinoza claims that all of Being, what he calls "Nature" is ***necessary*** both in its cause (God) and in its effects (the universe); *what* things are, therefore, is only a mode of the one divine essence, and therefore equally necessary.

... *MM,* with its inductive methodology, can find no verifiable necessity outside of conceptual tautologies, but does not deny the possibility.

*   For Spinoza there is a difference between "**extension**" as an "attribute" of "God" and the "**matter**" which makes up modal bodies. On this basis he denies the possibility that "God" has a body even while affirming "extension" as an essential feature of the divine substance. Extension is a *divine idea* ... and for that reason is creative.

... *MM* rejects all such notions of the causative reality of ideas. Ideas for *MM* are functions of human consciousness, the products of an organism's tool of survival.

dualism and idealism

Spinoza's characterization of modal bodies as inert and passive[252] and **opposed** to *thought,* repeats the prejudice inherited from perennial western dualism. It *describes all* reality (modal and substantial) as having two *irreducible* qualities, thought and extension. He attempts to avoid dualism by claiming to derive both from a prior and more fundamental **monism** ("God"), within which he has embedded an *ideal* materiality he calls "extension." The inconsistencies here, in my opinion, reflect his acceptance of the then unchallenged premises of Cartesian phenomenology, *viz.,* that

---

[252] "Body" is in motion or at rest *only in response to* another body in motion or at rest ... he doesn't speak of any *conatus* or self initiated movement that belongs to "body." *Conatus* is a feature of essence alone.

reality is made up of mind and matter — assumed to be entirely different modes of being.

But his system, inconsistent as it may be, claims to escape dualism on two counts. First he asserts that since "thought and extension" are the two attributes of "God," then anything that necessarily proceeds from the divine substance are necessarily constructed of the same two qualities, thought and extension, which he calls at the level of modes, "ideas and body." This is true of every modal expression; in other words **everything**, no matter how pre-life, inert-looking and lifeless, is comprised of thought and extension. *They must be,* because they are all modes of one substance, "God" whose infinite attributes are thought and extension.

And second, Spinoza declares that there is no "mind-body" dichotomy because *each term refers the same thing* seen from two different perspectives. Thought and extension are like the two dimensions of a rectangle, length and width, that are mutually irreducible yet each accurately describes the entire rectangle.

The "mind-body" "problem" debated today is, in my opinion, an unnecessary continuation of the same prejudicial Cartesian assumption that Spinoza rejected. Against claims that such rejections are gratuitous, Spinoza must be granted this: no matter how distinct "body and ideas" might appear on the phenomenal plane, both are *metaphysically grounded* in the absolute indivisible unity of the divine substance.

But problems remain: Spinoza denies that "God" has a "body" even though he says "God is an extended thing." What can that mean? Extension, or as we may say, **the materiality of "God,"** is an *idea* that has been taken for the divine "reality" itself;[253] it has creative power. Because divine materiality is "God," all bodies emanate from and remain *in "God"* even though God has no body. *There is no supernaturality* because nothing is "supernatural" to "God." But "body" is only modal. God can't have a body because nothing inert and passive can be true of God. How something can be so opposed to the divine substance and still emanate from it "necessarily" is not explained.

If Spinoza is to be accused of error in this complex attempt to reconcile *dualist phenomena and monist metaphysics* it must be laid squarely at the feet of his **idealism.** Like Plato and the scholastics he imputes causative, creative reality to *abstract ideas.*

---

[253] Henry E.Allison, *Benedict de Spinoza*, New Haven, 1987, pp.52-3

ethics

The posthumously published *Ethics* (1677) is considered Spinoza's greatest work, the product of his mature thinking. It is often cited and receives extended commentary because of the metaphysical groundwork laid at the beginning. But an exclusive focus on Part I and II of the *Ethics* can cause us to miss the fundamental thrust of Spinoza's work which deal with **ethics.** Spinoza elaborated his metaphysics for the sole purpose of providing the premises for *logically deducing,* in a manner reminiscent of Euclid's geometry, a view of human behavior that was the main object of his work.

Spinoza proposed to **deduce** *all human inclination and subsequent behavior from a human nature* whose sole interest was **survival** — a con-natural impulse he called **conatus sese conservandi,** *the drive to self-preservation.* The effect was to eliminate once and for all any notion of human behavior grounded in "eternal and transcendent principles," the "immortal spiritual soul," the "imitation of divine perfection," the "good of others," ... or of human defect due to "original sin," the "wiles of Satan," or any factor that implied the existence of a another world. By defining the human being as "**body-with-*conatus***," and deducing body, in turn, from the infinite **materiality** of "God" himself, Spinoza does away with all divisions imagined to exist between "God" and the material universe and therefore *all supernaturality.* For me, this highlights the fact that supernaturality is a corollary of dualism. For it is the matter-spirit division alone that provides the metaphysical justification for imagining the existence of a world different from the one that our bodies live in ... to which we must be *elevated* by the mechanical rituals of Catholicism.

By defining "mind" as "the idea of the body," Spinoza has subordinated the former to the latter. The priority belongs to **body,** *not to thought.* While this appears to be inconsistent with his statements about extension and thought in Part I, in my opinion it more accurately reflects reality. At any rate, it agrees with the view presented in *MM* — *consciousness is a property and function of material energy.*

appendix 9
molecular evolution table
elaborated by chemical engineer, Libb Thims,
http://www.humanthermodynamics.com/Evolution-Table.html

BYA= billions of years ago; MYA= millions of years ago

Element Count	Molecular Formula	Name	Formation Date
N/A	"Bang Plasma"	16 Fundamental Particles	13.7 BYA
N/A	u_2d+e (two "up" quarks, one "down" quark and gluons + electron) = proton	Subatomic	
1	H_2	Hydrogen	13.3 BYA
2	H_2O	Water	
3	CH_4O	Methanol	4.6 BYA
4	CH_4ON_2	Urea	
5	$C_{10}H_{12}O_6N_5P$	RNA	4.4 BYA
6	$C_{21}H_{36}O_{16}N_7P_3S$	Coenzyme A	
7	$C_4H_3O_2NS_2NaI$	mol-7	
8	$C_{20}H_{22}O_7N_2SNaIPt$	mol-8	
9			
10	$C_{E5}H_{E5}O_{E4}N_{E4}P_{E2}S_{E2}Ca_{E2}K_{E2}Cl_{E2}Na_{E2}$	Intermediate	4.2 BYA
11			
12	$C_{E7}H_{E7}O_{E6}N_{E6}P_{E4}S_{E4}Ca_{E4}K_{E3}Cl_{E3}Na_{E3}Mg_{E2}Fe_{E2}$	Pro-bacteria	4.0 BYA

13			
14			
15	$C_{E10}H_{E10}O_{E10}N_{E9}P_{E8}S_{E8}Ca_{E8}K_{E6}Cl_{E6}Na_{E6}Mg_{E6}Fe_{E5}Si_{E4}Mn_{E2}Co_{E2}$	Prokaryote	3.85 BYA
16			
17			
18	$C_{E16}H_{E16}O_{E16}N_{E15}P_{E14}S_{E14}Ca_{E14}K_{E12}Cl_{E12}Na_{E12}Mg_{E12}Fe_{E11}F_{E11}Si_{E10}Cu_{E9}Mn_{E8}Se_{E8}Co_{E7}$	Pre Aquatic Worm	2.6 BYA
19			
20			
21			
22	$C_{E22}H_{E22}O_{E22}N_{E21}P_{E20}S_{E19}Ca_{E20}K_{E18}Cl_{E18}Na_{E18}Mg_{E18}Fe_{E17}F_{E17}Zn_{E16}Si_{E16}Cu_{E15}I_{E14}Mn_{E14}Se_{E14}Mo_{E13}Co_{E13}V_{E12}$	Fish	0.7 BYA
23		Reptile	350 MYA
24	$C_{E26}H_{E26}O_{E26}N_{E25}P_{E24}S_{E23}Ca_{E24}K_{E22}Cl_{E22}Na_{E22}Mg_{E22}Fe_{E21}F_{E21}Zn_{E20}Si_{E20}Cu_{E19}B_{E19}I_{E18}Mn_{E18}Se_{E18}Cr_{E18}Mo_{E17}Co_{E17}V_{E16}$	Old World Monkey	30 MYA
25		A. Afarensis	5 MYA
26	$C_{E27}H_{E27}O_{E27}N_{E26}P_{E25}S_{E24}Ca_{E25}K_{E24}Cl_{E24}Na_{E24}Mg_{E24}Fe_{E23}F_{E23}Zn_{E22}Si_{E22}Cu_{E21}B_{E21}I_{E20}Sn_{E20}Mn_{E20}Se_{E20}Cr_{E20}Ni_{E20}Mo_{E19}Co_{E19}V_{E18}$	Human	0.2 MYA

appendix 10
lamarckian teleology in evolution

a. intelligence and social learning

We suggested in the text that there may be something like a "social Lamarckism" functioning in the future evolution of the human species. It might also be possible to make a case that a "Lamarckian" teleology was *already* involved in the past development of human intelligence. The thesis is the following: Intelligence, at whatever level, reveals itself to be a socializing force in evolutionary development, precisely because it represents not only the increased ability to learn from the environment (and thus *adapt* to *new* conditions, i.e., evolve by "natural" selection), but also an increased ability to learn from the group (and thus accommodate to *pre-existing social* conditions). The ability to learn from others means that as intelligence grows through successive speciation, reproductive success is being tied, increasingly, to the social conditions created by the group, even if they are less efficient for dealing with a particular environmental adversity than what some individual might come up with alone. As technology exercises more control over environmental conditions, intelligence is increasingly selected by the group for the purposes of *group cohesion* and the integration of the individual into society. The natural environment, though never entirely neutralized as a selection factor, is *increasingly* secondary. Response to the environment, in fact, is mediated by the group.

Therefore, with the introduction of socially conditioned intelligence, there is a new influence, a *social intentionality* mixing in with natural environmental factors determining which individuals will survive and which won't, and which behavior will be rewarded. The "hard wiring" taking place over time in the human genome tends to be put there by human society and reflects perennially perceived social needs. If a species has a number of potential ways of responding to a change in environment, the response that will dominate will *not* be the individual variation that shows the most efficient adjustment to the physical *environment alone*, but rather that variation that preserves *just enough* adjustment to the environmental conditions as to insure group survivability while in fact *prioritizing adjustment to the demands of the group* which is of much greater importance both for the physical survival and reproductive success of the individual. Adaptation in this case is "caused" by two "selective" factors, the "natural" environment and the "social" or "artificial" environment, with the emphasis being placed *increasingly* on the latter. This means that as speciation evolves "upwards" (meaning more social and more intelligent), the species themselves have an increasingly influential (though never exclusively determinative) role in the direction that development will take. It seems that *the ability to learn socially* had to have been a factor in the selection of intelligence itself. Some have conjectured that this was the driving force behind the emergence of specifically human intelligence 2.4 million years ago in the Pleistocene era.[254]

A "Lamarckian social teleology" is similarly visible in the extended parental care given to offspring not yet capable of independence. It is a fact that the period of childhood dependency came to be longer and longer as species complexified. How "natural" is humankind itself whose very existence depended upon these social commitments on the part of parents to sustain a childhood that was *not naturally self-sustainable,* i.e., it was not adapted to its natural environment? This is applicable, proportionately speaking, to all species that give extended care to their young. All such care speaks to a social intervention skewing "natural selection" and *preventing* the elimination of the "unfit." If natural selection — meaning selection by physical environmental conditions alone without teleological input from the organisms and their society — were the exclusive force in the formation of species, no animals with young that were not immediately capable of surviving would have evolved. None of the primates, for example, would have emerged. The correlation between greater intelligence and longer periods of childhood dependency means that the causal agent that has selected for intelligence is *society*, in other words, there is a teleological influence coming from the evolving species themselves.

It should be noted in passing that *society* here does not necessarily equate to organizationally complex behavioral interactions, but rather the depth of relational bonds and dependency. Highly elaborate interactive

[254] Peter Richerson and Robert Boyd, *The Pleistocene and the Origins of Human Culture*, Version 1.1. February, 1998. For presentation at 5th Biannual Symposium on the Science of Behavior: Behavior, Evolution, and Culture. February 23, 1998, University of Guadalajara, Mexico.

colonies of social insects, for example, is the result of natural selection working on instinctive algorithms without much intelligence or relational depth among individuals. Human sociality is quite intense, even in the absence of survival interaction.

This process reaches a current peak in humankind, where the modification of the natural environment by the manipulation of the social environment means that even the natural factors that might have had a direct say in what traits are selected, are themselves modified, i.e., "pre-selected," by human beings. The suggested Lamarckian factor is not a simplistic "internal desire directing development," but rather a more complex "internal desire embedded in community needs *modifying the environment*," which then selects within the conditions made available by a modified society.

It is not difficult to extrapolate back in time beyond humanity and see that the capacity for social learning, which is present in an ever increasing degree as animal species emerged in the evolutionary process, had to have been at work (analogously and to a lesser degree of course, affecting conditions throughout evolutionary history and therefore the directions speciation would take.

b. the "hardwiring" of learned characteristics

Stephen Pinker in his 1994 book *The Language Instinct*[255] makes this remarkable statement: "... natural selection can take skills that were acquired with effort and uncertainty and hardwire them into the brain."[256]

He cites his own "chapter 8" as support for this statement without giving a specific reference. In chapter 8, however, we find the following paragraph:

> Even when a trait starts off as a product of learning, it does not have to remain so. Evolutionary theory, supported by computer simulations, has shown that when an environment is stable, there is a selective pressure for learned abilities to become increasingly innate. That is because if an ability is innate, it can be deployed earlier in the lifespan of the creature, and there is less of a chance that an unlucky creature will miss out on the experience that would have been necessary to teach it.[257]

---

[255] First published by William Morrow and Co. in 1994. The page references are to the Harper Perennial Classics paperback printed in 2000.
[256] Ibid. p.377
[257] Ibid., p.244

However much I may agree with that position, we have to admit that these statements fly in the face of classic Darwinian theory which eschews simplistic Lamarckism. What can Pinker be talking about here except the "inheritance of acquired characteristics" which is the very definition of un-nuanced Lamarckism. It would hardly count as a rejection of Lamarck if Darwinians were simply denying that inheritance occurs in one generation, but have no objection to the same phenomenon occurring over a series of generations. Such a limited denial would amount to no objection at all.

I have a problem with Pinker here, not because of the content (the hardwiring, which I agree with) but because of his explanation and justification. First, he claims proof for his statement by saying an "evolutionary *theory* has shown" it. A *theory* cannot "show" anything. Secondly he informs us there are "computer simulations" that support this. What are they? We are not told. Finally he "explains" that the drift into "hardwiring" is "because" (and proceeds once again to justify theory on the basis of a teleology in nature itself:) This "because" is a product of Pinker's imagination.

The point is the hardwiring he speaks of is the conversion of a trait into a hereditary character. That conversion is the core issue of the controversies in evolutionary theory — the central problem of heredity that separated Darwin from Lamarck. If classic "reductionist" Darwinism is going to be consistently applied, the conversion must be explained in *non-teleological* terms. Nature does not *do* anything. It must be explained as the result of *survival selection* alone.

Frankly, I have no problem with survival selection. I merely differ from those who claim that all evolutionary selection *must be natural* (meaning not purposely chosen), and that all variations must be random. Not so. Teleology, choice, may be involved in both the variation and the selection process ... by the *intelligent* group. Remember that it was by reflecting on the mechanisms of the *intentional selection* exercised by people in the breeding of animals for desirable traits that Darwin came to understand how evolution could work by *natural selection.* The inviolable principle is *selection,* which leaves the organism *passive* in its own development. And if society modifies the natural environment in the exercise of that selection pressure, then we must recognize that it is *society* that is doing the selecting, and such selection is teleological; it is purposeful. This is what I mean by **social Lamarckism**.

This "intentional effect" need not be limited to the rationally teleological choices of human society. They may very well be true for animal traits as well. Richard Dawkins, in *The Ancestor's Tale,* Houghton Mifflin, Boston, 2004, p.199 in commenting on an example taken from Darwin says:

> I suspect that major new departures in evolution often start ... with a piece of lateral thinking by an individual who discovers a new and useful trick and learns to perfect it. If the habit is then imitated by others, including perhaps, the individual's own children, there will be a new selection pressure set up. Natural selection will favour genetic predispositions to be good at learning the new trick and much will follow. I suspect that something like this is how 'instinctive' feeding habits such as tree-hammering in woodpeckers, and mollusk-smashing in thrushes and sea otters, got their start. *

It appears undeniable that *learning* and its associated *intentionality*, admittedly mediated through "selection" (natural or purposeful), may very well be responsible for the variations that account for the divergence of species. This means *equally* that neither is there need to posit an outside "Planner-God" to explain divergence, nor must we accede to the absurd claims of the random reductionists that there is "no teleology whatsoever" functioning in evolution.

appendix 11
the social intelligence hypothesis

(this is an extended quote taken from Kerstin Dautenhahn (2001) "The Narrative Intelligence Hypothesis: In Search of the Transactional Format of Narratives in Humans and Other Social Animals," Proceedings CT2001, The Fourth International Conference on Cognitive Technology: INSTRUMENTS OF MIND (CT2001), LNAI 2117, pp. 248-266, Springer Verlag.)

... the *Social Intelligence Hypothesis* (SIH), sometimes also called Machiavellian Intelligence Hypothesis or *Social Brain Hypothesis*, suggests that the primate brain and primate intelligence evolved in adaptation to the need to operate in large groups where structure and cohesion of the group required a detailed understanding of group members, cf. [Byrne & Whiten 88], [Whiten & Byrne 97], [Byrne 97].

... there are two interesting aspects to human sociality: it served as an evolutionary constraint which led to an increase of brain size in primates, which in return led to an increased capacity to further develop social complexity.

A detailed analysis by Dunbar and his collaborators gives evidence (e.g. [Dunbar 92,93,98] and other publications) that the size of a cohesive social group in primates is a function of relative neocortical volume (volume of neocortex divided by volume of the rest of the brain). This evidence supports the argument that social complexity played a causal role in primate brain evolution, namely that in order to manage larger groups, bigger brains are needed to provide the required 'information processing capacity'.

The *Narrative Intelligence Hypothesis* [Dautenhahn 99a] interprets such evidence from the ontogeny of human language in the context of primate evolution: it proposes that the evolutionary origin of communicating in stories co-evolved with increasing social dynamics among our human ancestors, in particular the necessity to communicate about third-party relationships (which in humans seems to reach the highest degree of sophistication among all apes, cf. gossip and manipulation, [Sinderman 82]). According to the NIH human narrative intelligence might have evolved because the structure and format of narrative is particularly suited to communicate about the social world.

Narrative might be the 'natural' format for encoding and transmitting meaningful, socially relevant information (e.g. emotions and intentions of group members). Humans use language to learn about other people and third-party relationships, to manipulate people, to bond with people, to break up or reinforce relationships. Studies show that people spend about 60% of conversations on gossiping about relationships and personal experiences [Dunbar 93]. Thus, a primary role of language might have been to communicate about social issues, to get to know other group members, to synchronize group behavior, to preserve group cohesion.

Thus, stories are primarily dealing with people and their intentions ...

By mastering interpersonal timing and sharing of topics in such dyadic interactions children's transition from primary to pragmatic communication is supported. It seems that imitation games with caretakers play an important part in a child's development of the concept of 'person' and [Meltzoff

& Gopnik 93; Meltzoff & Moore 99], a major milestone in the development of social cognition in humans.

Data by Bruner and Feldman (1993) and others indicates that children with autism seem to have difficulty in organizing their experiences in a narrative format, as well as a difficulty in understanding the narrative format that people usually use to regulate their interactions. People with autism tend to *describe* rather than to *narrate*, lacking the specific causal, temporal, and intentional pragmatic markers needed for storymaking. A preliminary study reported by Bruner and Feldman (1993) with highfunctioning children with autism indicated that although they understood stories (gave appropriate answers when asked questions during the reading of the story), they showed great difficulty in retelling the story, i.e. *composing* a story based on what they know. The stories they told preserved many events and the correct sequence, but lacked the proper emphasis on *important and meaningful events*, events that motivated the plot and the actors.

These findings provide an empirical link between behavioral innovation, social learning capacities, and brain size in mammals. The ability to learn from others, invent new behaviors, and use tools may have played pivotal roles in primate brain evolution. [258]

The most likely evolutionary scenario is that where complex skills contribute to fitness, sociability and/or the capacity for socially biased learning increase, innovative abilities (i.e., intelligence) follow indirectly. We suggest that the evolution of high intelligence will often be a byproduct of selection on abilities for socially biased learning that are needed to acquire important skills, and hence that high intelligence should be most common in sociable rather than solitary organisms. Evidence for increased sociability during hominid evolution is consistent with this new hypothesis. [259]

---

[258] **Social intelligence, Innovation, and Enhanced Brain Size in Primates. Reader SM, Laland KN.** Department of Zoology, University of Cambridge, High Street, Madingley, Cambridge CB3 8AA, United Kingdom. Proc Natl Acad Sci U S A. 2002 Apr 2;99(7):4436-41. Epub 2002 Mar 12. PMID: 11891325 [PubMed - indexed for MEDLINE]

[259] **van Schaik CP, Pradhan GR. A model for tool-use traditions in primates: implications for the coevolution of culture and cognition.** Department of Biological Anthropology and Anatomy, Duke University, Box 90383, Durham, NC 27708-0383, USA. gauri@ces.iisc.ernet.in PMID: 12799157 [PubMed - indexed for MEDLINE]

appendix 12
schopenhauer's "will" & *material energy*

There is a substantial similarity between Arthur Schopenhauer's concept of "will" and the dynamism I have been calling *matter's energy.* Schopenhauer appears unique because of his determined choice of the word "will" to describe what is for him, as for me, a fundamental force that is operative in all things. His insistence on the word, however, derives from the context of his thought; it is not gratuitous or arbitrary.

I believe Schopenhauer's predilection for the word "will" comes from two sources. The first is the **Idealism** of the 19th century to which he, like Hegel, was a defining contributor.

Schopenhauer's philosophy was idealist, but in stark contrast to Hegelian enthusiasm, it was deeply pessimistic, and he was virulently opposed to Hegel whom he considered a charlatan. Schopenhauer's basic thought, contained in *The World as Will and Idea,* was first made public in a series of lectures given at the University of Berlin in 1820. If the term "Will" evokes something of a Transcendent Mind, however blind and impersonal it might be, it's because Schopenhauer was convinced of the metaphysical primacy of "spirit" (mind, thought) over matter. Thought creates and controls matter, and the manifest desire, "will," of all things to live was the reflection of their Mental "nature."

The second reason why Schopenhauer chose the word "will" for this singular and all-explanatory force, is his **methodology**. His approach was based on examining the phenomenon of *human will,* experienced internally and therefore undeniably by every human being, which he then extrapolated to all of reality. "Will" was an experience that served as starting point and theme of his philosophy. What we experience in us as "will" is the same force that resides at the core of all reality of whatever kind. It is therefore the fundamental dynamism of the Universe. It is the essence of being like Spinoza's *conatus.* I basically agree with that extrapolation and find it a valid procedure, although there are significant differences.

"Will" makes "being" to be a dynamism, not a "thing." "Things," according to Schopenhauer's Idealism, are the phenomenal objectifications of "will" created by a Spiritual Force and reconstructed according to embedded "Platonic" categories by the human mind. The realities we think we know are merely appearances that our minds have constructed. What's really there are different densities or intensities of "will," which, apart from our mental constructs, is unknowable. This follows Kant's *a priorism,* and by it Schopenhaur dares to have identified Kant's *noumena,* the allegedly unknowable real realities that lay beyond phenomena which we know.

This extension of "will" to all of reality is the keynote of Schopenhauer's philosophy. In its transition to universal application, however, one would expect "will" would have lost its psychological connotation and become indistinguishable from blind, physical "energy" or the "force of life," as it does in my approach. A force like gravity, for instance, was a manifestation of "will" for Schopenhauer and an example of "will's" impersonality. The word "will," however, continues to suggest a psychological dimension. Hence Copleston rightly asks, "Would not 'Force' or 'Energy' be a more appropriate term, especially as the so-called Will, when considered in itself, is said to be 'without knowledge and merely a blind incessant impulse',[260] 'an endless striving'[261]?"[262]

But I believe there is a reason for Schopenhauer's insistence on the word. For Schopenhauer, in a key transition to his moral philosophy, condemns "will" as selfish and predatory. Such a characterization shows that he never had *any intention* of shedding the psychological and moral connotations of the term. The energy of being, in this sense of "willful," is selfish and malevolent; it is the justification for his pessimistic view of life and the world-rejecting *asceticism he proposes* as a response. One suspects that Schopenhauer was working backwards from a preferred conclusion.

It's exactly here that we can see the similarities and differences between Schopenhauer's "will" and the concept of *matter's energy* proposed in *MM.* Empirically I believe we are speaking of exactly the same phenomenon. Reality is a thirst, a dynamism, not "things." We even agree that the inner human feeling of the "will to live," the **conatus,** the *drive to survive* that we experience at the very core of our personal selves is the manifestation in us of the very same force operating at the core of all reality. I

---

[260] Copelston's fn. #23 *The World as Will and Idea,* tr. Haldane & Kemp I, p. 354
[261] Copleston's fn. #24, ibid. p.213
[262] Frederick Copleston, *A History of Philosophy,* vol.7 part II, Doubleday, Image, p.37

also agree that this dynamism generates a primitive, rudimentary teleology, a purpose: *survival.*

We differ, however, in regard to how we characterize this energy. Schopenhauer's Idealism has no explanation for "will," though the term evokes the psychological emanation of a Transcendental "Mind." It's the essence of reality, but it is blind and senseless. For my part I believe in our times we have discovered that "energy" is the physical constitution of matter itself. It's a physical property, not a demonic emanation. In itself it does not have a moral dimension. It has an obvious ability to combine with itself in ways that create identity and thence time and space. I describe the inclination to such perdurance metaphorically as a *self-embrace.* It is survival — the driving force of evolution. This *energy,* then, is the basis of a cosmo-ontology that embraces everything. And since I do not, as Schopenhauer does, condemn survival as "evil," I have no difficulty in identifying it as the source of our sense of the Sacred.

Schopenhauer sees "will's" survival function as blindly *selfish* and therefore ultimately evil and destructive. (How something so "destructive" could have been responsible for our unspeakably beautiful cosmos is a paradox he either did not notice or chose to ignore — *despite his central emphasis on aesthetics.*) He believes our only salvation is to transcend this selfish energy, conquer it, negate the "will to live" by an heroic asceticism that stills the clamor of the self and awakens altruism. In contrast, I believe that *matter's energy* is the basis of all creative development of whatever kind. It is the engine of evolution. It is the ultimate source of the transparency of consciousness. Yes, it is originally blind and self-directed. But it has been responsible for the emergence of ever-increasing cognitive complexity in life forms, along with an increasing participation of the evolved species themselves in future evolution eventuating currently in our human consciousness with its capacities for contemplation and transcendence over selfishness. This is as much a development of *matter's energy* as any other. The fact that we have participated fully in the creative process — that what was once a blind and groping force to live, by our own directing hand has become in us a capability for the selfless stewardship of all things — does not cancel out the original potential coming from *matter's energy.* Far from it. If the renunciation of selfishness is considered "sacred" by Schopenhauer, the energy responsible for it must also be sacred. And that energy, by Schopenhauer's own definitions, must also be "will."

Schopenhauer extols asceticism. I believe we can see still operating in the background of this predilection the ancient distinctions that have defined good and evil in western thought since the days of the Greeks. Altruism, based on dispassionate contemplation (the ability to regard things in themselves and apart from our selfish interest in them), and sustained by ascetical discipline, is the key that he believes separated good from evil (and humankind from the animals). Traditionally this criterion was associated with the distinction between spirit and matter because only "spiritual beings" were believed capable of altruism. For the Idealists that distinction more than held, it became paramount, and Schopenhauer was an Idealist.

The demolition by Hume and Kant of any rational proof of God's existence and character drove Kant in a second step to re-establish "God and morality" based on an analysis of the inclinations of conscience, moral obligation and altruism. As an avowed disciple of Kant, it's no surprise to find Schopenhauer grounding what he calls "holiness" in the same ethical soil as Kant, even though Schopenhauer denies the existence of "God." The only criterion for "the sacred," as far as he is concerned, is the human recognition of the transcendent superiority of *altruistic morality* based on contemplation and renunciation. Thus, "the sacred" remains grounded in the moral aspirations of the human species and nowhere else.

Schopenhauer considers "will" selfish. I believe, in contrast, that *matter's energy* is a neutral force whose accomplishments reveal its *potentiality*, not its "final cause." *Matter's energy,* by exploring the possibilities open to it, *on its own power and initiative* over eons of time, has eventuated in the evolutionary emergence of what Schopenhauer would call "sacredness" at the current apex of the process displayed in the potential behavior of the human species. It is the potential negation of self and the development of altruism based on the human capacity for *aesthetic contemplation*.

He highlights this contemplative and aesthetic function and sees it as an *escape* from the world of slavish desire. He may have noticed the internal contradiction lurking just below the surface in this conception, because in an attempt to locate contemplation in a region apart from "will," he identifies it with Platonic Ideas as the objectifications of "Will." I believe this is a contrived idealist explanation that causes me to agree with Copleston who flatly declares "it is something which I do not profess to understand." [263]

I believe with Schopenhauer that contemplative cognition (and its associated altruism) is a by-product of our conscious survival apparatus. We agree that contemplation was discovered and cultivated by the human species. It is a creative development, the result of humanity participating with full engagement in its own evolution, utilizing the potential embedded in *matter's energy*. (See the text, chapter 7). That we have ourselves done this, far from derogating from the sacred potential in *matter's energy*, confirms it. This energy not only does not resist, it shows itself to be entirely amenable and connatural with the creative transcendence of co-existent contemplation.

We can see illustrated in this interplay a key perspective that dogmatic atheists like Schopenhauer seem to have great difficulty in understanding or accepting. He cannot admit the force of life is *sacred*, because he does not see a *direct and unmediated* causal connection between the operations of that force and the "sacred" behavior — renunciation, contemplation and altruism — that he cherishes so highly. So he pejoratively calls that force "will" and condemns it as selfish and evil. This reaction suggests a superficial and altogether simplistic reading of the old axiom *ex nihilo, nihil fit.* ("What exists in the effect had to have been present in the cause.") In this case it is quite evident there is a "sacred" development *mediated through the creative inventiveness of the human species*, finding and elaborating what was there, dormant, as a potential in human consciousness. This is not sleight-of-hand. The potential was both truly there *and* truly dormant. Without human intervention, it would have remained dormant forever. And so the human being is truly a co-creator in the elaboration of the Sacred. The Sacred, therefore, can be said to be truly *a posteriori,* a development made possible only through the involvement of humanity.

It is "will" itself, therefore, (or what I call *matter's energy*) that ultimately accounts for creative transcendence. For if this energy is responsible for everything, then it is ultimately responsible for Schopenhauer's very call to "conquer the will to live." "Will" functions on both sides of the issue: the "will" that must be overcome (the potential), and the "will" that does the overcoming (the activator). If, therefore, "will" is operative everywhere

---

[263] Copleston, *History* ... vol 7

even in contemplation and "sacred" self-transcendence, how can it be condemned as evil?

Schopenhauer apparently never asked himself these questions. And I believe it was because he was locked into the traditional renunciations counseled by the world-denying Religions, East and West, that continued to dominate the values of the intellectual elites of his era. (Schopenhauer was conspicuous in his attempt to integrate the East Indian "religions" into his philosophy.) Such "holiness" as these religions espoused implied in western terms a disdain for the desires of life, a repetition of the ancient hostility toward the "flesh." Thus, in spite of his counter-cultural atheism and pessimism, Schopenhauer ended up promoting a program that was indistinguishable from the religious and moral traditions inherited from the ancient past.

appendix 13
bergson's *elan vital* and matter's energy

The first of the so-called process philosophers, Henri Bergson attempted to explain all of reality as a function of evolutionary creativity driven by a *vital impulse*. In many respects my views in this essay agree with Bergson's; but I also differ from him in major ways.

Fundamentally, Bergson is an idealist; I'm not. Like Hegel, he believes that all of reality is the expression of consciousness, spirit, and his philosophy in great measure is a search for the clues scattered throughout the universe that support his interpretation. He sees evidence of the creative primacy of spirit literally everywhere, in all things and in all the processes of nature. Hence the *elan vital* — the vital impulse.

Even the existence of matter itself is given an idealist explanation. Bergson likens matter to the elemental building blocks of a poem: the words, phrases, syllables, no one of which in itself, is in the least poetic. The poem, he says, is not constructed by the dead elements. The elements are drawn into living existence by their place in the poem (*Creative*

Evolution, p.209f, 236-251).   The very existence of the elements represents a *distension,* i.e., a relaxing of the psychic *tension* which is the poem, a kind of *unmaking.* This *distension,* he says, results in *extension,* spatiality, the primary characteristic of matter.   He is so unapologetic about this idealist view of matter that he openly enlists no less than Plotinus himself in support. In a long footnote (p.210) Bergson refers to Enn. *IV,* iii, 9-10 and *III,* vi, 17-18 to show that Plotinus considered matter to be the ultimate *enfeeblement* (a notion he associates with his own *"distension,"* p.212) of the divine self-thinking, the "tail-end," as it were, of the emanation process. Matter was the last shadowy residue of the light of divine thought (i.e., "spirit") at the edge of oblivion.

Some may point out that Bergson himself denied that his position was "idealist." But he did so only on the basis of what he called the "block thinking" of both the empiricists and the idealists (*CE,* p.240). By "block thinking" he meant that they believed reality was a block of "things" passive to the action of another "thing." In his view, neither Idealist nor Empiricist recognized *change* to be the essence of reality. They did not think of duration, process, as real ... nor that reality was the result of a process and always in process toward other results. Evidently Bergson believed that he had achieved an integrated view that braided matter and spirit in a new way that transcended the inherited "block" dualisms of western philosophy. He seems unaware that the primacy he accords to spirit in the *process* that generates matter (thus defining matter as an expression of spirit) has placed him squarely in the camp of the Idealists, and, therefore, quite obviously, the dualists as well.

In contrast, I make no determinations about the origin of the matter of our universe, because for me matter is the primordial substance and there is no other. There is no explanation "beyond" matter. Whether the universe of matter contains its origins within itself, or in a "matter" outside itself, as Gregory of Nyssa speculated, is irrelevant to me because I do not have "two kinds of reality" to reconcile. There is only matter. There is no "spirit." Therefore in either case matter remains exactly what it is — *existence itself.* It's "nature and character" will not change with a change in etiology, since any hypothetical currently invisible ultimate source for the actual *material energy* that we see must itself be a *material energy* in some other appearance or "location." "A" will still equal "A." Existence, being, is one thing and one thing only: *matter's energy.*

So, while Bergson can lament, as I also do, the false conception that imagines "statically ready-made material particles juxtaposed to one another and, also statically, an external cause which plasters upon them a skillfully contrived organization" (p.249), he does so in the name of an integrated dynamic dualism that sees *matter* as the emergent product of creative *spirit*. My view, to the contrary, imagines one single reality, *matter's energy* in a constant process of achieving an *enduring self-embrace* through internal re-alignment and re-combination. To those who say they fail to see what is significantly different in these perspectives let me say very clearly and very directly: Bergson's view is dualist and idealist. It is pejoratively prejudicial to matter. He continues to subordinate matter to spirit *metaphysically*. In his view matter is generated by spirit. By ultimately denying matter independent reality, he undermines its rights *as matter*. Bergson is a neo-Platonist wolf garbed in the sheepskins of the professors of 19th century empirical science.

This metaphysical denigration of matter eventually, if not inevitably, results in Bergson's claim that matter is *opposed* to spirit. He identifies matter's role in the creative process as *resistant and inhibitive* and, though ultimately incapable of fully blocking the creative explorations of spirit (because matter, in his mind, is totally passive and incapable of any energy whatsoever, even negative energy), provides a retarding effect that he claims draws out the power of spirit to transcend its inertia. He describes the qualities of matter in terms that evoke its inertness, its impotence and its dependence on the vital power which is undeniably *other than itself*, something immaterial — "spirit." Indeed, in my opinion the main difference between Bergson's view and the "external cause plastering a contrived organization" on matter which he criticizes, resides in a further subordination of matter to spirit, now reaching to its very existence — though he doesn't explain how a spirit that creates matter can produce an offspring that is so opposed and resistant to its own spiritual initiatives.

So the *vital impulse*, central to Bergson's view of reality and to mine, belongs for him to another world even as he claims it is responsible for the developments in this one. At the root of his erroneous perspective, I believe, is the endemic western belief that matter is and can only be *inert*, dead, and entirely passive which Bergson accepts, consciously and intentionally or not. All energy of whatever kind, positive or negative, is denied to matter. And it is exactly this millennial assumption that has been so thoroughly demolished by modern science. Matter, we now know, not only *has* energy, it *is* energy. Even though Einstein's equations were roughly contemporaneous with the writing of *Creative Evolution*, there is no indication that Bergson was aware of them. In any case, the full realization of the significance of $E=mc^2$ did not emerge for most people until long after Bergson's time. To this day the overwhelming momentum of thousands of unchallenged years of the western belief in the passivity and obtuseness of matter still dominates our imagination if not our thinking. Once the inertness of matter is recognized to be seriously defective as an image of reality, the concomitant unchallenged assumption of the "necessary" existence of life-giving spirit will have lost its *raison d'être*. Without a passive, inert matter, all of reality must be re-interpreted; and along with it humankind's traditional role as the unique "spirit in the world," set apart from the world, which we took as permission to do with material reality whatever suited our fancy. Indeed, it was the unfortunate identification of "spirit" with human rationality that guaranteed that matter would always be judged as deficient in "being," justifying a "rational" management that often demanded the suppression and sometimes even the elimination of matter's natural character, properties and attributes.

Bergson's *vital impulse,* like Schopenhauer's *will,* does nothing to change the *status quo* — the Platonic paradigm.

appendix 14
chomsky and the "language faculty"

Noam Chomsky, in his *Managua Lectures* of 1986, published in *Language and Problems of Knowledge,* proposes that the structural limitations and requirements of a "language faculty" (a conjectured part of the brain responsible for an *a priori* "Universal Grammar [UG]") determines the grammatical structures of all languages. The "generative grammars" of specific languages take their guidelines from some element or another of the UG and operate only within its parameters.

Is there really a special and separate "language faculty" biologically embedded in the brain that creates, structures and limits language? Chomsky rejects the point of view he calls "environmentalist" which, he chides, is accepted as much without question as without evidence. Environmentalism, in the linguistic context, means that language is primarily a product of "nurture" not "nature;" it is learned, it's not innate. Chomsky says environmentalism cannot explain certain features of linguistic capability among humans, like early childhood grammatical accuracy. Only a "hard-wired," genetically programmed biological apparatus can account for language — and its derivation, *number* — as we know it.

What seems to bother Chomsky is the environmentalist claim that all language is "learned," and all facility with language is the result of habit. The environmental hypothesis would imply, he says, that it is theoretically possible to construct languages that exhibit radically different structures with radically different grammatical features from those that exist on earth today. He denies that to be a fact, and then tries to show by multiple examples that certain "principles of grammar" are "universal;" they are elements of a UG, and function in all extant languages, revealing a fixed biological etiology.

First, I am not convinced that the handful of examples he offers actually prove what he claims they do. Then, from another angle, the fact that language is the property of humankind alone would be consistent with the environmentalist theory; for a "biological" etiology as he espouses would expect to find language existing to varying degrees in species leading up to humankind. Such is not the case.

What's worrisome for me is that Chomsky's proposal involves the possible re-introduction of a *metaphysical discontinuity* implied by the existence of a special human faculty which does not share its structure, function or development with the more general processes affecting all other life forms made of *matter's energy*. The human being, in other words, is different from all other animals, including primates not only in degree but *in kind*. The existence of a special "language faculty" (which must be taken to mean a special "intelligence faculty") evokes Platonic "spiritism" and intellectual *a priorism*. Indeed, Chomsky admits his proposal is designed to answer what he calls "Plato's problem," viz., "where does our (apparently innate) knowledge come from?" Plato's answer, still operational in the time of Descartes, traditionally meant belief in the presence of a unique immaterial "immortal soul" in each individual, rendering the human species entirely distinct from the rest of the material and living universe. We will recognize that such a discontinuity, in fact, has not only been traditional in western culture, it has been its most prominent and influential feature and accounts for the anthropocentrism and the intense individualism that dominates our value system. I seriously doubt that our culture would be the same without the doctrine of the "immortal soul." It is arguable that the belief in the otherworldly origins and destiny of the human species has been responsible for the exploitative relationship to the earth so characteristic of our western culture. Our culture proposes that everything besides humankind is mere soulless matter, an alien substance available for whatever use or abuse we may decide. This is not insignificant.

In point of fact, however, Chomsky does not say that language requires the existence of an immaterial "soul." I believe we can avert any implied challenge to the integrity and continuity of the substrate, including the human species, by incorporating Chomsky's theories into our overall hypothesis about *matter's energy,* and showing that everything he would want from a "language faculty" is already present and available in the **general intelligence apparatus** which evolved under the same environmental conditions and material dynamism as the animals and, indeed, everything else on earth. In fact, since Chomsky's main objection to "environmentalism" revolves around the issue of a genetic coding for language, it seems that acknowledging the existence of such a coding within a broader biological context, and focused more generally on the "one and the many" that included the animals, should address his concerns as well as ours. This proposed solution would have recourse to neither "environmentalism" nor an exclusively human "language faculty" with the grammatical character that Chomsky proposes. I will try to explain what I mean in the paragraphs that follow.

language

Humans, just like the animals, form aggregated images — concepts. This function represents the conscious grasping of the fundamental structure of cosmic reality: the "one and the many," as we discussed in the text, chapter 5. But the labeling, the "word," language, that accompanies human conceptualization is unique and allows the resulting concepts to be organized on a new level. The use of these symbols is the basis of logic as well as grammar and mathematics, which are fundamentally the same thing. I believe the "computationally complex" rules of grammar claimed by Chomsky to explain the phenomenon to which he refers in *Language and Problems of Knowledge*[264] are simply examples of the hierarchical ordering of verbal symbols, as he admits on p.45: "The rules operate on expressions that are assigned a certain structure in terms of a **hierarchy of phrases** of various types." The requirement of the "R-Q rule" (his shorthand for turning complex statements into questions) is "to find the 'most prominent' occurrence" of a verb within a series that otherwise would seem unstructured or "flat."

The human biological apparatus, says Chomsky, demands that the rules be "structure dependent." I believe this is neither the unexpected arcane discovery that he claims it is, nor does it fly in the face of the traditional belief that language is elaborated along the lines of "the one and the many" that are common to all other activities of intelligence. Verbal symbol-making (language) may require a new biological apparatus like special vocal chords and mandible position, but I contend that *there is no need to fabricate a unique brain-based "language faculty" to account for the hierarchialization that characterizes "grammar."* I believe the *conceptualization* (the grasp of the one and the many) that is the basis of all language is created by the same mechanisms and the same brain structures utilized for every other human operation involving intelligence and is fundamentally similar to that of the animals.

This is an important issue: I am convinced there is no "special" human biological faculty of consciousness that is denied to the animals. For animals also *conceptualize* by aggregating a multiplicity of like experiences. They "hierarchialize" at the most primitive level, but it is still hierarchialization. *What the animals lack is the ability to use symbols — the word or label.* So, yes, language is unique to humans. And, of course, the power to verbalize represents a biological advance embedded in both the brain and the vocal apparatus. But it's important to emphasize that the uniqueness of language *does not consist in the ability to aggregate singularities,* (to grasp the one and the many), but rather in our ability to "label," and therefore to aggregate our own aggregations, *discretize* and thus abstract from individual imagery, creating a thoroughly communal and highly mani-

---

[264] Noam Chomsky, *Language and Problems of Knowledge, The Managua Lectures.* MIT Press, Cambridge, MA, 1988 (lectures delivered March, 1986).

pulable product, viz., *speech based on words* which are themselves capable of complex, multi-level hierarchialization. If the ability to universalize, i.e., to grasp "the one and the many" is survival-related, as I claim, then it's clear why this is a characteristic of the entire animal class. Its more sophisticated applications, on display as the human ability to "label," and then to manipulate — compare, organize, classify, include, exclude — labels, is simply a *symbolically managed* version of the inclusions and exclusions of the "one and the many" of which the animals have been capable for an immense amount of geologic time prior to the appearance of humankind.

A derived function like language, no matter how sophisticated it appears and transcendent its product, should be seen as a *subset* of the more fundamental ability to generalize experience. I believe by these criteria, language with its *symbolically guided* hierarchical structures reveals itself to be merely an example of the inclusions that define animal intelligence — the capacity to understand the "one and the many."

The only separate human biological "faculty" required, according to my view, is for making symbols, not for the more basic ability to universalize. The grammatical features that Chomsky identifies as "innate," may very well have followed Pinker's process of the "hardwiring of a learned response." Such a development in no way prejudices their subordination to the universal capacity to conceptualize that I claim is the common property of all living consciousness and the basis of grammar.

Homo sapiens is a rather late development on the hominid scene. Since all earlier versions of homo are now extinct, we have no way of identifying the various stages of the development of the language faculty which surely would have illuminated this question.

index

www.ingramcontent.com/pod-product-compliance
Lightning Source LLC
Chambersburg PA
CBHW031459270326
41930CB00006B/165